Joseph Fernandez

Henry's Outlines of Science

Third Edition

Joseph Fernandez

Henry's Outlines of Science
Third Edition

ISBN/EAN: 9783337034597

Printed in Europe, USA, Canada, Australia, Japan

Cover: Foto ©berggeist007 / pixelio.de

More available books at **www.hansebooks.com**

HENRY'S

OUTLINES OF SCIENCE.

BY

JOSEPH FERNANDEZ, B.A., LL.D.,

AUTHOR OF "HENRY'S SCHOOL GEOGRAPHY," "HENRY'S DICTA-
TION," "HENRY'S SCRIPTURE HISTORY," ETC., ETC.

THIRD EDITION.

LONDON:
CHARLES BEAN, 81, NEW NORTH ROAD, HOXTON, N
1872.

PREFACE.

The following series of lessons on the Elements of Science has been prepared to meet a want which has been strongly expressed to the Author by many of the multitude of teachers who have done him the honour to adopt his other educational works.

Without attempting a complete epitome of any of the sciences discussed, it is hoped that sufficient information is given to render the book a good introduction to a fuller study of the subjects.

Pains have been taken to direct attention as far as possible to what are called common things.

Numerous questions on the text have been added.

Eagle House, Tottenham,
January, 1871.

OUTLINES OF SCIENCE.

ASTRONOMY.

CHAPTER I.

Astronomy is the science which treats of the heavenly bodies; their sizes, motions, distances, periods, and the order in which they move.

The term Astronomy is derived from two Greek words, ἀστήρ (aster), a star, and νόμος (nomos), a law.

The Solar System consists of the planetary bodies with their satellites or attendant moons, which pass round Sol, the sun, and are influenced by its light and heat.

In the present system, which is called Pythagorean, from Pythagoras, an old Greek philosopher, who lived B.C. 550, the sun is the centre of all the other planets.

Fig. 1.

Ptolemaic System.—For many centuries the system of Pythagoras was not accepted, but men insisted that the earth was a flat surface, and that the sun went round it, or rather went over it, during the day, and having passed under it, came up again at the other side.

Ptolemy, a celebrated astronomer of Egypt, who lived at Alexandria in the second century, A.D. 130, placed the earth in the centre of the universe, and made all the other planets revolve round it.

Copernicus.—From the time of Ptolemy, until the sixteenth century, little progress was made in the science, but in A.D. 1500, Copernicus, a native of Thorn, in Prussia, began the study, and during the next thirteen years published works on the subject, in which he set aside the Ptolemaic System, and introduced that which is now generally received as the Copernican System, in which the sun is the centre, and all other planets revolve round it.

His opinions met with violent opposition, and he died in 1513, worn out with anxiety and excessive labour in the pursuit of his favourite study.

Tycho Brahe.—Tycho Brahe, who lived in the same century, and who was a native of Denmark, rejected the theories of Ptolemy and Copernicus, and maintained that the earth was in the centre, and that the sun revolved round the earth, while all the other planets similarly revolved round the sun. This cumbrous method was believed in until gradually argued away, and replaced by the Copernican, in the various works of Kepler, Galileo, Newton, Halley, Herschel, and other more recent astronomers.

Primary Planets.—In the Solar System all the planets which revolve round the sun are called Primary, and the number, as discovered hitherto, amounts to eight, besides the sun itself. These are as follows: *reckoning from the sun*—MERCURY, VENUS, EARTH, MARS, JUPITER, SATURN, URANUS, and NEPTUNE. These vary greatly in size, as well as in the diameter of their orbit or travelling path, and in distance from the sun. They may be known by their steady light and the absence of twinkling. Several of them have secondary planets or moons, called satellites or attendants, which pass round them in stated periods, just as our moon travels round the earth. While the Earth has *one* moon, Jupiter has *four*, Saturn has *eight*, Uranus has *six*, and probably more, and Neptune *one*, as far as is yet known.

Planetoids, or Little Planets.—Between the orbits of Mars and Jupiter there is a greater space in proportion, than between the other planets, and there are many small bodies, all of which have been discovered within the present century, and many during the past twenty years, since the size and power of the telescope have been so much increased. Before 1847, the only planetoids, or asteroids, as they are commonly called, known to astronomers, were Ceres, Pallas, Juno, and Vesta, which were discovered between 1801 and 1807.

QUESTIONS ON CHAPTER I.

Of what does Astronomy treat?
From what words is Astronomy derived?
Of what does the Solar System consist?
Who was Pythagoras? When did he live?
What was the peculiarity of the Ptolemaic System?
What name is given to the Solar System at present?
Who rejected the Copernican theory? Who accepted it?
Name the primary planets in their order from the sun?
What are the secondary planets? What is a satellite?
Where are the planetoids? Name the four largest?
When and where did the astronomer Ptolemy live?
How many moons or satellites has Jupiter?
How may the planets be distinguished from the stars?

CHAPTER II.

THE EARTH.

Form of the Earth.—The earth, being the planet

which we inhabit, is of most interest and importance to us. For many ages people thought that the world was flat, and the old writers of Greece and Rome speak of the ends of the world, and of the Ultima Thule, the most distant or last regions. Thales, who lived about B.C. 609, had a notion that the world was round.

We have many proofs of its *rotundity* or roundness. The proof so common in our books on geography is most familiar, which notices that a vessel on leaving the sea-shore is lost sight of by degrees, and that the main-topmast disappears last, which it would not do if the earth were flat, because it is easier to see the large hull of a ship, than the small spars at the top of it.

Again, if two ships at sea are one mile apart, the water rises eight inches above the level; if two miles apart, *four times* eight inches; if three miles apart, *nine* times eight inches; and so on, the height of the water increasing in similar proportion as the square of the distance increases.

Another more conclusive proof is that caused by a lunar eclipse, that is, an eclipse of Luna, the moon. When the earth comes exactly between the sun and the moon, the shadow of the earth is cast upon the lunar surface. This is always round, and it may easily be proved that only a round body can in all positions cast a circular shadow.

Diurnal Motion.—The earth turns round upon its own axis or centre from west to east, once in about every twenty-four hours; this motion causes the alternation of day and night. We do not notice this motion, because everything on the surface of the globe, and in the atmosphere, is carried along with it.

Because of this motion, the sun appears to rise in the east, and to move across the heavens in a certain path, which is called the ecliptic ; and the clusters of stars, called constellations, also appear to do so.

As all the works of God show such manifest wisdom, as well as adaptation, we may regard it as certain that bodies of such great size and at such vast distances, do not go round the earth. During twenty-four hours, nearly every part of the earth's surface comes opposite the sun. We therefore travel at the rate of one thousand miles an hour.

It is easy to understand from this diurnal motion, that while the people who are opposite the sun enjoy his noon-day or meridian heat, those who are on the opposite side of the world are in the darkness of midnight. This will account for the fact that when it is twelve o'clock noon at Greenwich, it is twelve midnight in Van Diemen's Land, seven o'clock at Philadelphia, and so on at intermediate places.

The earth is not exactly round, because a line drawn round its equator would be $26\frac{1}{2}$ miles longer than a similar line drawn round it from pole to pole. The equatorial diameter is therefore so much longer than the polar.

It is therefore called an oblate spheroid, which means that it is flattened a little at the poles like an orange. It is believed that the difference is caused by the rapid motion, which causes it to bulge at the equator, while it is a little diminished at the poles.

QUESTIONS ON CHAPTER II.

What was the opinion of Thales ? When did he live ?
Give three proofs of the rotundity of the earth.
What is meant by the diurnal motion of the earth ?
What is the result of the daily motion ?
What is the ecliptic ? What are constellations ?
When it is noon at Greenwich, what time is it in Australia ?
What is an oblate spheroid ? What makes the earth one ?
In which direction does the earth turn ?

Hence where does the sun appear to rise?
At what rate do we travel on the earth's surface?
Why is the Copernican theory most reasonable?

CHAPTER III.
ANNUAL MOTION.

Annual Motion.—Besides its motion on its own axis, the earth travels round the sun in 365 days, 5 hours, 48 minutes, and 49 seconds, or, as we usually say, in 365¼ days. This motion is the cause of the difference of the seasons, and also in the length of days and nights in various parts of the globe.

This motion is proved by observing the passage of the sun through the ecliptic, when it will be noticed that his position changes a little every day until, in the course of a year, he passes through all the constellations of the Zodiac.

Thus, in January he is in Capricorn, in March in Aries, while in June he is in Cancer, and in September in Libra, and during this annual motion the seasons are constantly changing.

To understand the cause of this, we must look at the figures which follow. If, as in Fig. 2, the axis of the earth were perpendicular, we should have equal day and night all over the earth, because the same half of the earth would always be twelve hours within the light of the sun. In this case the parts of the earth opposite the sun would be excessively hot, while the polar regions would be as cold in comparison—a man living on the parallel of latitude E, would always enjoy a temperate climate, and not be subjected to change of season.

Fig. 2.

The fact is, however, that the axis or pole of the earth is not perpendicular or upright, but inclined 23½ degrees, as in Fig. 3. In its path round the sun, the axis is always kept thus, and consequently all the parts of the earth are warmed and lighted in turns, in proportion as they receive the direct rays of light and heat. The people who live on the equator have always day and night of the same length, that is twelve hours; the people at the poles, if there be any, have a day six months long, and a night six months long, because each pole is within the sun's rays for that period of the year. The nearer the equator, the less is the difference between the length of the day and night, while the nearer the poles, or the higher the latitude, the greater is the difference in their length.

Fig. 3.

The Seasons.—These vary as the length of day and night varies. From Fig. 3, it will be seen that a person who lives at A will be on a parallel, which in the summer season of the year will be in the sun's light sixteen hours, and in the winter season only eight hours. Consequently that parallel will receive more heat in sixteen hours than it will part with in the following eight, and the heat increases as the days lengthen, and even for some time beyond. Thus July and August are usually hotter than June. In winter the opposite will be the case, and the cold of the long nights will not be removed by the warmth of the short days. Hence the old winter adage—

"As the day lengthens, the cold strengthens."

In the months of March and September the axis of the earth does not decline in either direction, and on those occasions the days and nights are equal all over the world. These are called the *vernal* and *autumnal equinoxes*.

Dimensions.—The *circumference* (or measure round) of the earth is found to be 24,907 miles, which is therefore the length of the equator, and of all other great circles. In Fig. 2, all lines drawn round the earth passing through the poles are great circles, and are called meridians. As these pass through the equator, they are used to measure *longitude*, east or west.

All circles which pass round the earth parallel to the equator, and either north or south of it, are called parallels of *latitude*, because by means of them altitude is measured, north or south.

The diameter of the earth from pole to pole is 7899 miles, and through from the equator to the opposite side is 7926 miles nearly.

Orbit and Distance from the Sun.—The orbit or path in which the earth travels round the sun, is in the form of an ellipse, or of a circle drawn out a little. The sun is in the position of one of the nodes of the ellipse, so that the earth is nearer the sun at one period of its course than at another.

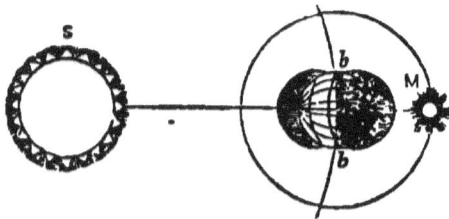

Fig. 4.

This happens in winter, when the world is nearest to the sun, but as it moves quicker, it does not appear to derive any increase of heat. The greatest distance of the earth from the sun is 93,000,000 miles, its least distance 90,000,000 miles. The mean or average distance is therefore 91,350,000 miles. Until recently it was said to be 95,000,000 miles.

QUESTIONS ON CHAPTER III.

What is the annual motion of the earth?

What is the result of this annual motion?
How can this annual motion be discovered?
What would happen if the earth's axis were upright?
What is its amount of inclination?
What is the length of day on the equator?
How are the seasons said to vary?
When is the effect of accumulated heat or cold felt most?
In what months are the vernal and autumnal equinoxes?
What is the circumference of the earth? Its diameter?
What is a meridian? and what a parallel?
What is the earth's orbit? What is its form?
When is the sun nearest to the earth?
What is the greatest distance between them? And the average?

CHAPTER IV.

Longitude.—As the circumference of the earth is over 24,000 miles, it follows that the sun's light passes over it at a rate of more than 1000 miles an hour. The equator, which is a great circle, is divided into 360 degrees, this divided by 24 gives 15, the number of degrees passed over in every hour.

The longitude of a place, is the difference between the clock time at Greenwich, and the clock time at any other place, reckoning fifteen degrees per hour, or one degree for every four minutes.

Thus the captain of a ship who finds the sun at its meridian when his Greenwich-corrected chronometer is at ten o'clock, will know that his longitude is two hours east, or thirty degrees east. Should he find the sun reaching its meridian when his chronometer marks two o'clock, he will similarly know that his longitude is two hours, or thirty degrees west.

Chronometers.—It is of the utmost importance that these chronometers should keep exact time, otherwise finding the longitude by this process would be uncertain, and consequently dangerous to the navigator.

Chronometers were brought to great perfection by John Harrison, of London, who received a reward of £20,000 for some which enabled captains to find out their position within thirty miles, after a long voyage. This was in 1787, and they have since been much improved.

Latitude.—To find latitude, which is the distance north or south of the equator, the observer must fix on some of the heavenly bodies and find their altitude or height above the horizon, when they come to their meridian. As the times of the meridian of the sun and all the heavenly bodies, with many other particulars, are given in nautical almanacs, the observer, having found the longitude also, is able to see his exact position on the surface of the ocean.

Tides.—There is an ebb and flow of all the waters on the globe *twice* in about twenty-five hours (twenty-four hours fifty minutes). This is caused by the attraction of the sun and moon. It was discovered by Sir Isaac Newton, that substances are kept in their places while rolling through space at the rate of 1000 miles an hour, by a force called the attraction of gravitation; and all the heavenly bodies in the Solar System are influenced by this force, so as to be kept in their orbits.

As the moon passes round the earth in her orbit, the tides appear to follow, as if attracted by her. They are always highest on the side next the moon, but there are also tides on the side opposite the moon, because the force of gravity decreases as the square of the

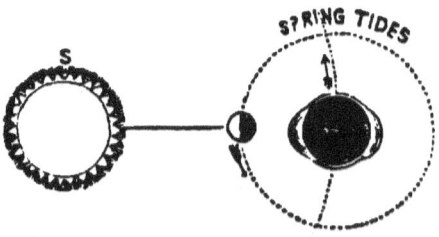

Fig. 5.

ASTRONOMY. 15

distance increases. The denser parts of the earth are drawn by the attraction of the moon, and the water being drawn with less force, causes the tide on the opposite surface.

The tides are therefore always highest when the sun and moon are in opposition, and when in conjunction, or on the same side.

When they are in *quadrature*, or attracting at right angles, the tides are lower. So they are highest when the moon is new, and when she is in her third quarter.

As all the water on the globe comes under the influence of the moon twice during the twenty-four hours, it causes two tides during that time.

High water, therefore, occurs about fifty minutes later one day than that of the previous day.

The tides are not of equal height in all places. In the Bristol Channel the average rise is sixty feet. In the Bay of Fundy, and in St. Malo Bay, it often rises to a height of one hundred feet, while at places more sheltered it does not rise above ten or fifteen feet.

Eclipses.—When the moon, in its monthly revolution round the earth, comes exactly between that planet and the sun, it intercepts part of the sun's light. This is called solar eclipse, or a concealment, and the earth, moon, and sun, are in the same straight line. Fig. 6. The solid body of the moon hides a part of the sun from the inhabitants of the earth who happen to be on the same side. Sometimes the surface of the sun is entirely hidden, this is a *total* eclipse; at other times it is covered except a small ring round the

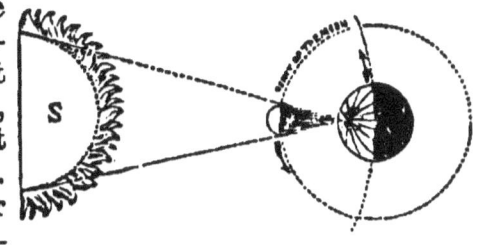

Fig. 6.

outside of the moon, this is called an *annular* eclipse, from annulus, a ring; at other times only a part of the sun is hidden, which is called a partial eclipse. A solar eclipse can happen only when the moon is either very old or very new.

A lunar eclipse can only take place when the moon is full, or nearly so, because then only does the earth lie between the sun and moon, so as to allow its shadow to fall upon the moon. Fig. 7.

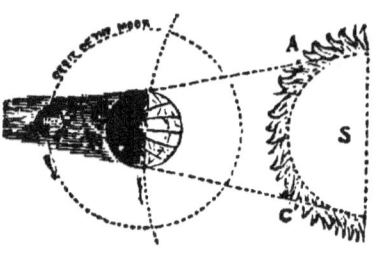

Fig. 7.

Both solar and lunar eclipses would happen every month, if the orbit of the moon were in the same plane with that of the earth. This is not so, but it makes an angle of about $5\frac{1}{4}$ degrees, with the plane of the orbit of the earth, and crosses it in two points called nodes.

To understand this, take two hoops, one a little less than the other, place the smaller inside the larger and open them a little; you have then a picture of the two orbits and their nodes. It is only when the moon passes through these nodes that an eclipse can occur, and this happens about four times a year, so that we have two Solar and two Lunar eclipses on the average.

An eclipse of the sun begins at the western side of his disc, but an eclipse of the moon begins at the eastern side of that planet.

By lunar eclipses we prove that the earth is a round body, that the earth is greater than the moon, and that the sun is greater than either. We also learn how long it takes the moon to complete its circles round the earth.

QUESTIONS ON CHAPTER IV.

What is longitude? What is the circumference of the earth? How many degrees does the sun pass over in one hour?

What is a great circle? How is time kept at sea?
If a captain finds the sun in his meridian at two o'clock by his chronometer, what will be his longitude? If at nine o'clock A.M.?
Who perfected the chronometer? What proof was given of it?
What is latitude? What help is needed to find it?
How often do the tides ebb and flow? What do they seem to follow?
Where are the tides highest? When are they highest?
In what places do very high tides occur? To what height?
What is an eclipse? When does a solar eclipse happen?
What is an annular eclipse?
When only can a lunar eclipse occur?
Why does not a lunar eclipse happen every month?
On which side does a solar eclipse occur?
How often do eclipses happen in the year?
What do we learn from the earth's shadow on the moon?

CHAPTER V.

THE MOON.

Form and Size.—The moon is a globe similar to the earth, but much smaller, as its diameter is only about one-fourth that of the earth. It is distant from us about 240,000 miles. The difference in the distance of the sun and moon from our earth will account for the little apparent difference in their size, for while the diameter of the sun is really 400 times that of the moon, one seems nearly as large as the other. The earth is forty-nine times larger than the moon.

The moon always shows the same face to us, though it is certain that she revolves on her own axis slowly, and so as to complete the revolution in about a month.

From its comparative nearness to us, we know something of its general appearance. It seems to be covered over with hollow craters or cups, which are

2—3

believed to be empty volcanoes. It does not appear to have any atmosphere, like that of the earth and the sun, nor is there any sign of water.

Half of it is always in the light, just as the earth is, and no doubt the earth appears to the inhabitants of the moon, if there be any, just as the moon appears to us, and it affords to them the same help with its reflected light.

Phases of the Moon.—By Fig. 8, it will be easily seen how the moon passes through its various phases or appearances in its passage round the earth. When the moon is at A, it is between the sun and earth, and its dark side is turned to us. It is only when it reaches B, that we see a small part of the lighted surface in form of a crescent or horn, which we call new moon. At C we see the half moon, and its lighted surface continues to widen until it reaches D, when it is full moon, after which it grows gradually less towards E, until it is lost sight of again at A. We may often see the full moon during the daytime, but it is always on the side of the heavens opposite to the sun, that is, always at D.

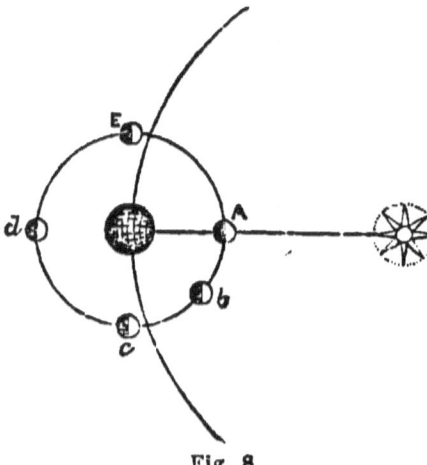

Fig. 8.

Harvest Moon, &c.—When speaking of the high tides, it will be remembered that they were stated to occur about fifty minutes later each day. The moon rises, as it is called, about so much later every day, because when the world has gone its daily round, the same meridian will not be opposite the moon, because the moon has been also moving on in its monthly course. By a singular concurrence of events, the moon at one season rises at shorter intervals for several days together, than during the rest of the year. This occurs at the end of August and part of September, when the fruits of the earth are being gathered, and it is therefore called the Harvest Moon. It is a great blessing, because the light is very much prolonged by the moonlight commencing before the daylight has ended.

The Winter Moon.—In winter the northern hemisphere is turned away from the sun (See Fig. 3), and leans toward the full moon; the moon has therefore a greater altitude as compared with the earth, and consequently shines longer and brighter. By this means we have more moonlight in winter than in summer, because in summer the northern hemisphere is turned toward the sun, and the moon does not reach so high an altitude above the horizon.

Easter Moon.—It has long been settled among Christians that " Easter Sunday is the first Sunday after the full moon which happens on or next after the 21st of March." It is on this account that Easter occurs later on one year than another.

THE SUN.

By observing the spots on the disc or face of the sun, astronomers have ascertained that it revolves on its own axis from west to east in about twenty-five days.

Its diameter is about 107 times that of the earth, and its bulk in comparison as 1,250,000 to *one*, so that his bulk or volume is many times greater than all the other planets put together.

To give some idea of the distance of the sun from

us, it is said that if a cannon ball were fired at an even velocity of sixteen feet in a second, it would take nearly ten years to reach that great luminary.

Amount of Heat.—Many speculations have been made about the probable cause of the sun's heat. Some have guessed it to be a ball of fire; but if it were of consumable matter, it must have long since burnt itself out. Others have imagined its heat to be caused by incessant striking of meteoric bodies on its surface; but the real cause is quite unsettled.

Although the earth receives a comparatively small portion of its heat, it is estimated that the amount received yearly would melt 100 feet thick of ice, or boil sixty-six feet deep of water all over the globe.

Some philosophers say that "all force is heat" from the steam which George Stephenson called "the sun's heat bottled up" to the strength in man which is needed for the smallest amount of labour.

Latent Heat.—Much of the heat of the sun is called *latent*, or hidden, because it is only developed by friction, or some other force.

There is latent heat in ice; for it requires nearly as much heat to melt the ice to ice-cold water, as it does afterwards to raise the water to the boiling point, 212°. The familiar case of making buttons or other pieces of metal hot by rubbing them, is well-known to every school-boy, and that is an example of the development of latent heat. By this process of friction pieces of dry wood are kindled by savages, and the axles of wheels become red-hot if left ungreased.

QUESTIONS ON CHAPTER V.

What is the size of the moon as compared with the earth?
Give some description of the moon's appearance.
Where is the moon, when its dark side is turned to us?

Where is the moon when we see her at the full?
How much later each day does the moon rise?
What other phenomenon is similar to this?
When does the harvest moon occur?
When have we the most moonlight? Why?
What relation has Easter to the moon?
How do we know that the sun turns on his axis?
What is his distance from the earth?
Why cannot the sun be a body of fire?
What did Stephenson call steam?
What is latent heat? Where is it found?
Give instances of developing latent heat.

CHAPTER VI.

THE PLANETS.

Mercury, ☿.—When the astronomers of old watched the heavenly bodies, they noticed that some of the large ones did not always keep the same positions as they found the fixed stars to do. They therefore called them planets or wanderers.

The planet Mercury is the smallest, and is nearest the sun in the Solar System, being distant about 35,000,000 miles, and going round him in eighty-eight days.

The nearer the sun, the faster a planet moves, so that the rate of travelling of Mercury through space is 105,000 miles an hour, while that of the earth is 66,000. The diameter of Mercury is about 3,000 miles, and he is so close to the sun that he can never be seen except in twilight, that is just before sunrise, or just after sunset.

Transits of Mercury, or passages across the disc of the sun, which occur about once in ten years on an average, show that he is an opaque body like the earth, and has phases like those of the moon, which vary with his position.

Venus, ♀.—This planet is 66,000,000 miles from the sun, and has a diameter of 7896 miles. She goes round the sun in 225 days, and seems to be sometimes larger than at other times. The axis of Venus is more inclined than that of the earth, and therefore there must be a greater variety in the seasons. Her transit only happens once in about sixty years. We see much more of this beautiful planet than of Mercury, as a morning or evening star. When west of the sun she appears a morning star, and when east of him an evening star.

Next to Venus is the Earth, ⊕, which has been more fully noticed elsewhere.

Mars, ♂.—Beyond the earth, that is, outside of it from the sun, the next planet is Mars. His distance from the sun is about 139,000,000 miles, his diameter only 4,175 miles, about half that of the earth, and his year about 687 days, or not quite two of our years. The diameter of Mars appears greater to us because his distance from us is only 48,000,000 miles, and we are able to see him with greater ease. His daily revolution is known by a large and very remarkable spot on his disc. He may be readily distinguished by his red appearance, from which the ancients named him after Mars, their god of war.

The Asteroids, or little stars, are most of them very small, and are contained in the great space of the Solar System, between Mars and Jupiter. They amount to eighty in number, and the largest is only about $\frac{1}{700}$ part of our moon, while the smallest is not large enough to be measured. The reasonable opinion of astronomers is, that they once formed one planet, which by some extraordinary explosion or convulsion has been blown to pieces.

Jupiter, ♃.—This planet is distant from the sun 490,000,000 miles, and is 1,240 times larger than the earth; his diameter is about one-tenth of that of the sun, while Jupiter's is eleven times that of the earth. His year or revolution round the sun takes nearly twelve years, but his day is not quite ten hours. This very rapid diurnal rotation causes the equatorial regions to bulge out, so that there is believed to be a difference of 6,000 miles between the polar and the equatorial diameters. The axis of Jupiter is perpendicular to his orbit, and not inclined, as in most other of the planets.

Jupiter has four satellites or moons of large size, which are easy of observation. The largest of these is half the size of the earth, that is about equal in size to Mars, two others are one-third, and the smallest one-fourth the size of the earth. He has also dark streaks across his disc called belts, which vary in size at different times, and are supposed to be his atmosphere in his various conditions of quiet or excitement.

The moons of Jupiter afford valuable help in measuring the velocity or swiftness of light. It is found that when Jupiter is on the same side of the sun that the earth is, astronomers see his moons emerge from their eclipses seventeen minutes earlier than when the earth and Jupiter are on opposite sides of the sun. Light therefore must travel across the intermediate space of nearly 190,000,000 miles during that time, which gives the wonderful velocity of 192,000 miles in a second.

This remarkable circumstance first attracted the notice of the famous Galileo, the Italian astronomer, who was imprisoned by the Inquisition because he asserted that the world moved.

QUESTIONS ON CHAPTER VI.

What planet is nearest to the sun?
Why were some heavenly bodies called planets?
At what rate must Mercury travel?
When only can he be seen? and why?
How do we know that Mercury is an opaque body?
What is the distance of Venus from the sun?
What is the length of her year?
Why must there be a great variety in her seasons?
When does Venus appear a morning star?
Between which planets is the earth?
What is the distance of Mars from the sun?
What is his distance from the earth? His year?
How may Mars be easily known?
What are found between Mars and Jupiter?
How many are known to exist?
What is the size of Jupiter as compared with the earth?
The length of his day? And of his year?
What is peculiar about his diameter?
How many moons has Jupiter?
At what rate does light travel? How is this known?

CHAPTER VII.

THE PLANETS—*Continued.*

Saturn, ♄.—This planet is nearly as large as Jupiter, having a diameter of 73,800 miles. His mean distance from the sun is 871,000,000 miles, his year is nearly as long as thirty of ours, while his day is only about 10½ hours.

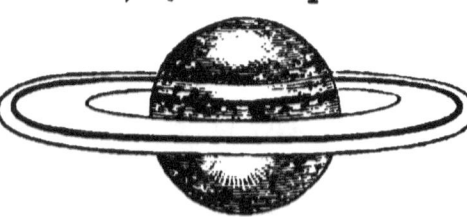

Fig. 9.

Saturn has eight moons, besides two, if not three rings, which can only be seen from the earth, when he is elevated above the orbit of that planet. These rings are believed to revolve round the planet, and the largest of the moons turns on its own axis, as our moon does.

Uranus, ♅, also called Herschel.—The next planet, called Uranus, was discovered by Dr. Herschel, in 1781, and formerly called Georgium Sidus, or George's Star, in honour of King George III., who was the great friend and patron of Herschel.

This planet is 1,800,000,000 of miles from the sun. His year is equal to eighty-four of ours, and he is about eighty times greater than the earth. Uranus has several moons, but their number is not yet accurately known. They revolve in a different direction round the primary planet, from all other planets and moons—that is, around the *polar* instead of the *equatorial* circumference.

Neptune, ♆.—This planet, which is the most distant of all at present known to astronomers, was discovered by Mr. Adams, in 1845, and shortly afterwards by M. Le Verrier, a French astronomer. The new planet is 2,473,000,000 of miles from the sun, and his year is equal to 165 of ours. He is thought to be a little larger than Uranus. At present only one of his satellites has been discovered, but there are doubtless several others.

Lately, the French astronomers have spoken of another new planet only 14,000,000 of miles from the sun, but at present the fact of its existence is not ascertained. It is to be named Vulcan, if it is proved to be a planet.

Comets. — These are hairy-looking luminous bodies connected with the Solar System, about which very little is known, but whose appearance in former times threw the people of the world into a great fright. They move in what are called eccentric ellipses, and their periods of revolution, or of re-appearance, are very little known.

Though more than one hundred have been noticed, only a few have appeared more than once, and these at very long intervals, being at one time comparatively close to the sun, and then passing beyond the reach of the most powerful telescopes. Nothing is known of their composition, but they cannot be very solid, as the fixed stars and planets have been seen through the gauze-like tail and through the nucleus or head of the comet.

Some approach the sun rapidly, and others at a slow rate; some have several tails, but most of them have only one. Only three have settled revolutions and re-appearances; those are Halley's, Encke's, and Biela's comets. Halley's comet returns every seventy-five years, Encke's in about three and a-third years, and Biela's in about six years and eight months.

Fixed Stars.—These are heavenly bodies so far beyond the Solar System, that they cannot be magnified by the most powerful telescopes. Unlike the planets, which shine with a borrowed or reflected light, each star shines by its own light, and since they are at so vast a distance, they are supposed to be each a centre of some planetary system, similar to ours.

Though not more than 16,000 stars can be seen with the unassisted eye, there are supposed to be vast numbers of them. 150,000 have been registered, and Sir W. Herschel said that 50,000 passed his great telescope in one hour. Some are in pairs, others in threes, which revolve round each other, and in some parts of the heavens they are so numerous as to form star-clouds, or nebulæ, like the Milky Way.

The Milky Way.—This is the great luminous band which stretches every evening all across the sky.

ASTRONOMY. 27

It is divided in one part of its course, sending off a kind of branch, which joins the main body again.

This remarkable belt, when examined through a telescope, is found to consist of millions of stars, which are scattered like dust over the vast heavens.

QUESTIONS ON CHAPTER VII.

What is the diameter of Saturn?
Give the length of Saturn's day and year.
How is the planet easily known?
How many moons? and rings?
What planet is outside of Saturn?
By whom was he discovered? What was he called?
What is the year of Uranus? His distance from the sun?
What is peculiar about the moons of Uranus?
When and by whom was Neptune discovered?
What is his distance from the sun? Length of his year?
Where do French astronomers think there is another planet?
By what name is it to be known when discovered?
What are comets? How do they move?
How many have settled appearances? When?
How can a fixed star be distinguished from a planet?
What are the Nebulæ supposed to be?
Describe the Milky Way.
Of what is it found to consist?

BOTANY.

CHAPTER VIII.

BOTANY (from Βοτανη, a plant) is the science which treats of the structure and properties of the vegetable kingdom; from the largest tree, to the tiny moss or seaweed.

Plants, as well as animals, are provided with organs which are necessary to their existence, and which answer to the various means with which the latter are supplied. They choose food and reject it, they breathe and perspire, they sleep, and have a circulating fluid.

Plants differ from animals in this respect, that they feed on inorganic substances, such as minerals, water, and air, while animals are supported chiefly by organic substances; so that plants live on minerals, and animals are nourished by plants and vegetables.

Structural Botany is that part of the science which describes the elementary organs of the plant, such as the *root*, the *stem* or *axis*, the *leaves*, the *flower*, the *fruit*, and the *seed*.

Of these the root, stem, and leaves feed the plant, and are called organs of *nutrition*, while the flower, fruit, and seed are called organs of *reproduction*, because by their agency the new plant is generally produced. We say generally, because plants are often propagated by slips and cuttings, or by dividing the root, as well as by seeds.

Physiological Botany treats of the *internal* structure of the plants, the *tissues* of which they are composed, their *food*, mode of assimilating it, organs of respiration and circulation.

Other branches of botany are *Descriptive* and *Systematic;* our business is chiefly with the structure and physiology of plants, about which botanists are agreed.

Parts of a Plant.—The *root*, or radicula (1), is the organ which chiefly supplies the plant with food. This is either water, or something held in solution by water; the root is therefore generally buried in the earth, where the moisture is found. The root will strike downward, even though the seed may be intentionally placed the wrong way up.

The root never produces buds, and in small plants is usually of a white or pale colour, because it is hidden from the light.

Fig. 10.

On examining carefully the root of the common buttercup, or heart's-ease, for example, we find that the roots grow gradually smaller like threads, and at the end of each rootlet there is a small tuft.

3—3

If this be cut off, split up, and placed under a microscope, or a good glass, it will be seen that the end of it is covered with a little sheath, and the part which is thus covered or protected is the growing point of the root, which is constantly extending itself in search of food, through the rough hard soil into fresh places. This sheath is fed from the inside by the little growing point which it protects.

The stem or **caudex** (2) begins where the proper root ends. It is usually coloured green; and bears foliage leaves on each side of it of numerous forms and variously situated or arranged, sometimes opposite to each other, sometimes alternate. From the lower leaf-joints, called *nodes*, smaller stems usually spring, and the stem gradually grows more slender until it reaches the flower. The flower-stalk is called a *peduncle* (3), at the end of which is the *receptacle* (4), of a colour similar to the stem.

Inside the receptacle are other green leaves called the sepals, and together the *calyx*, (4) or flower-cup, and within them coloured leaves, which form its chief beauty, and are called the *corolla* (5). The calyx and corolla are composed of several separate parts, which vary in number according to the class to which the flower belongs. Inside the flower are found a number of small stems or stamens (6), on the end of which there is to be seen a kind of dust called pollen, and usually in the centre of the stamens, there is one or more longer than the rest, and somewhat different in form, which is called the pistil (7). This is hollow, and at its base is situated the seed-vessel (8).

Before the seed can be formed, the pollen must be passed from the stamens down the centre of the pistil, where it gradually expands and assumes its proper form, the leaves of both calyx and corolla fall off, and the seed-vessel having ripened in the heat of the sun, bursts and allows the ripe seed to escape on to the ground, unless preserved by artificial means.

Thus we have the root, stem, leaves, peduncle, receptacle, calyx, corolla, stamens, pistil, and seed-vessel, all of which perform an important part in the production of the fruit.

Varieties of Root.—These are (1) the *tap* or *pivot* root, which is long, taper, and simple, like the carrot, parsnip, and beet-root, or is globular like the turnip and bulbous ranunculus.

(2). **Branched** roots, in which, added to one principal root, there are others branching from it, as in most large trees and shrubs.

(3). **Fibrous** roots, consisting of long fibres, some very thin and fine, such as wheat, barley, and the other grasses. Roots vary somewhat according to the soil or position in which they are found, and it is remarkable that nearly all become fibrous when exposed to the action of running water.

(4). **Bulbous roots**, such as the onion, hyacinth, lily, &c.

There are some varieties which do not belong to any of the above. For example, the misletoe, and many lichens, which are called *parasitical* plants, because they fix their roots into other plants, instead of the ground, and appear to live on their juices.

There are also some roots which are called *adventitious*. These form creeping stems, which give off root fibres at the nodes, and which extend in this manner over a large space. Familiar examples of this are to be found in the common garden mint, and in the campanula.

In perennial herbaceous plants like the dahlia, the root thickens in the ground, and so becomes a reservoir of nutritive matter, which is thereby stored up during winter until required to nourish the young shoots of the coming spring.

As proper roots never give off stems, we must remember that the potato and other similar plants, from the eyes of which come the new shoots, are not *roots*, but only thickened portions of the stem, the potato root being quite separate.

QUESTIONS ON CHAPTER VIII.

What is Botany? How are plants like animals?
In what respect do they differ from them?
What is structural botany?
Name the organs of nutrition and of reproduction.
Of what does physiological botany treat?
How is a plant chiefly fed? What is said of the root?
What is found at the end of the root? How is it covered?
Name the other parts of the plant.
What are the various kinds of root?
Name some fibrous root. Some bulbous.
What are parasitical plants?
What are adventitious roots? Name some.
What is the advantage of the thick dahlia root?
Is the potato a root? What is it?

CHAPTER IX.

The Leaf and its Functions.—Leaves, as said elsewhere, assume very many forms, in different plants. Sometimes the leaf-stalk is wanting, and then the broad leaf, or lamina, adheres closely to the stem, and is called *sessile*, as in woodmint, sowthistle, and others. In other cases there are several produced from points forming a ring on the stem, which is called a *whorl* or *verticil*. Examples of this are seen in goose-grass, and mare's-tail. Others are *pinnate*, like the rose leaf, *entire* like laurel and lilac, *serrated* or notched like the nettle, *sinuated* like the oak-leaf, &c.

Whatever their form, they all have the same duty to perform. As the root is the organ of absorption, by which nourishment may be said to be sucked up from the soil, so the stem is the conducting medium of that nourishment, while the leaves are organs of *transpiration, assimilation,* and *respiration.*

Transpiration.—If fresh leaves be soaked in water, in the summer, and placed under glass in the sun, there will be a quantity of oxygen evolved, which can be collected and experimented upon; this giving out of oxygen, which has been received probably in the form of carbonic acid, is called *transpiration.*

Assimilation.—To receive it as carbonic acid, and to separate the oxygen so as to evolve it, as just mentioned, using the carbon for growth and nourishment, shows a function of the leaf which is called *assimilation.*

Respiration.—It is also known that the green parts of plants, in the night, or in dark places, absorb oxygen, and give off carbonic acid, and this process of drawing in oxygen is called *respiration* or breathing. Hence it is considered unhealthy to keep plants in a sleeping-room during the night, because carbonic acid in excess is injurious to human life.

Analysis.—By burning a plant, and analysing the ashes, it is found to contain various minerals, the chief of which are sulphur, soda, potash, phosphorus, lime, silex, and others, in small quantities, of course, and not all in the same plant, but sufficient to show that these substances have been absorbed by the roots. Silex, which is the basis of flint, is found in the stalks of wheat, bamboo, and all other grass plants, and in the teak, and other hard woods.

Food of Plants.—The organic matter of mould, consisting of decayed vegetables, &c., is continually forming carbonic acid. This is intended to be absorbed by plants with water, which is the chief medium through which they are fed. They also absorb nitrogen, and some kinds, if planted in pure sand, will increase in their per centage of nitrogen, which must therefore be derived from the atmosphere.

Those plants which have the largest leaves are found to be less exhaustive and injurious to the ground, as they derive so much of their nourishment from the atmosphere. This is the case with turnip, beet-root, and other large-leaved plants.

Air necessary.—Plants, like animals, must have air. If either a plant or seed be placed in the exhausted receiver of an air-pump, it will not grow, even though the mould in which it is planted is of the richest and most suitable character. No other gases besides those which form air, in their present combination, are suited to nourish plants; most of them are highly injurious, and cyanogen utterly destructive. Both leaves and roots require air, hence the advantage of frequently ploughing or otherwise disturbing the soil.

Fruit.—Until fruit is nearly ripe, it feeds on the atmosphere, as the leaves do. The young apple, when hard and almost tasteless, contains only woody fibre and starch, but as it grows it absorbs oxygen in *definite* proportions, and contains in succession tartaric acid, malic acid, and subsequently sugar.

Light.—This is necessary to the plant, but injurious to the seed, which germinates best when concealed from light. Sugar is also formed more rapidly in beet-roots and other plants, when the root is partly hidden by large leaves. Celery roots, if uncovered, would be bitter instead of sweet, and contain much less sugar. One result of abundant sunlight and heat is seen in the fact that the sweetest and most aromatic plants are found in warm countries.

Design.—It is especially worthy of notice, as a wise design in creation, that plants live upon an element, or part of it, which is injurious to man, namely, carbonic acid, and that having, as just shown, separated the carbon and used it, it transpires or breathes out the oxygen, which gives life and energy to man and other animals.

Irritability.—One peculiar property of leaves is seen in their power of closing or of moving, under the influence of weather, sunlight, atmospheric pressure, or the touch of any foreign object. Many plants close in the night, or, as Linnæus said, "they sleep." The leaves of the Acacia and many others, droop as night approaches, and revive when morning appears.

More striking examples are those of the Dionœa and the Mimosa pudica, or sensitive plant, the leaves of which close in pairs, the instant that plant is touched, until, if the touches be repeated, the whole leaf droops as if dying, and only revives after some time has passed. The Dionœa, or Venus fly-trap, a North American plant, has leaves which close suddenly when an insect lights upon the stiff prickly hairs with which their insides are covered, and kill the prisoner. The sun-dew, common in English marshes, has a similar property, but in a less degree.

Plants require *sun-light* to flourish, so that if put in a cellar with a small opening, the leaves will all turn their upper sides toward the opening, and the stalks will grow longer as if by stretching. The leaves of most plants fall off in cold climates; those which are evergreen, such as fir, pine, larch, and others, have peculiar fluids in their bark or skin, which defend them from the weather.

Those which lose their leaves are called *deciduous*,

and some which lose their leaves in cold climates, retain them in their native climes, such as the orange, oleander, and others.

Design may also be seen in the cotyledons or first leaves of a plant, which are usually larger and more fleshy than later leaves. They serve as protectors to the little germ, until it is strong enough to do without them, when they fade or die off the little plant. Familiar examples are seen in garden beans and peas.

QUESTIONS ON CHAPTER IX.

What is the function of the leaf?
Name some principal kinds of leaf.
What example is given of *transpiration*?
What is called *assimilation* in plants?
What is exhaled by plants during the night?
What minerals are commonly found in plants?
What articles form their principal food?
Prove that they receive nourishment from air.
What happens to a plant in an air-pump?
What is the advantage of ploughing the earth?
What are the components of a hard apple?
What does it absorb as it ripens?
How is a plant affected by light?
How do plants favourably affect the atmosphere?
What is irritability? Give some examples.
What more do plants require to flourish?
Distinguish between deciduous and evergreen.
What is remarkable about the cotyledons?

CHAPTER X.

The Bud.—The bud consists of a collection of the rudiments or commencement of leaves, surrounding a central vital point. It may be a leaf-bud, or a flower-bud; leaf-buds are seen in autumn, as they are produced in the axilla or centre of the leaf, and therefore appear when the leaf falls off. They are generally covered with thick leaves or scales, called *perules*, which protect them from the weather, and sometimes they are also coated with a resinous matter which may be best seen in the leaf-bud of the horse-chestnut.

Hairs and Glands.—Many plants, especially those which grow in exposed situations, have numerous hairs scattered over the stem and leaves. These are of two kinds, namely, the *lymphatics*, which are believed to regulate the escape of moisture by evaporation, and the *secreting* hairs or glands, which contain fluids peculiar to the plant.

Secretion.—A familiar example of the secreting gland is the hair on the stinging nettle, which, when touched, pierces the skin, and a sharp acrid juice flows into the wound from a little bag at the base of each hair, which causes the tingling or smarting pain.

It is believed that the perfume of plants and flowers arises from the secretions contained in the glands, which are most powerful when crushed.

The Flower.—This, as we have shown, is composed of several distinct portions, and is the most important part of a plant, as it is the seed producer. The flowering, or *inflorescence*, is various in form in different flowers. Sometimes, as in the peony, one flower is produced on one stem, and then ends; this is called *solitary* and *terminal*.

When all the buds on an axis become flowers, as in the hyacinth, it is called a *raceme*. If they are also *sessile*, as in corn, lavender, saintfoin, and many other plants, they form a *spike*, so that an ear of corn is a spike; the catkin of the nut, filbert, walnut, and other trees, are *pendent* because they hang down. When many flowers are in a bunch, as in the carrot, hemlock, celery, &c., each bunch is called an *umbel*, and the bunches on one peduncle are called a compound umbel—such plants are called umbelliferous.

The Stamens.—These are said to be the male organs of a plant. At the top of a thin filament is a small hollow case or sack, which is called an *anther;* this anther is filled with many minute globules called the *pollen,* which contains a fluid. Shortly after the expansion of the flower, the anther opens, and the pollen is scattered on the *pistil,* which is the female organ. This process is essential to the perfecting of the seed, and has sometimes to be performed by the hand, as in the cultivation of the cucumber.

The form of the stamens and pistil, with the anthers, may be plainly seen in all the lily tribe. Moreover, as these stamens have been made the basis of a system of botany, they are especially worthy of notice. High cultivation is very injurious to the productiveness of plants, as it often converts the stamens into petals, and makes the flower double. The anther then disappears, and no seeds are produced. So we never have seed from these double blossoms, or flowers.

The Pistil.—This usually consists of three parts, the *ovarium,* or seed-vessel, at the base, the *style,* and the *stigma.* When the pollen-fluid enters the stigma, it is conveyed into the ovarium, which is variously shaped, and often contains numerous cells or cavities, like the poppy and passion-flower. Sometimes the pistil is hidden as in the violet, and in other cases the anthers and pistil will have the same style or stem.

When the anthers have deposited the pollen on the pistil, their function ceases, and they, with the corolla, die off; but the pistil and ovarium continue to enlarge and ripen until the seeds arrive at maturity, and fall, or are collected to be resown.

Fruit, in botany, means the ripened ovarium; it

BOTANY. 39

may be a pea-shell, an apple, a strawberry, a marshmallow cheese, or a simple flower-seed like the sweet-william, or any ordinary annual. The ripe seed-vessel is called the *pericarp*, the outer skin is the *epicarp*, the pulp is the *sarcocarp*, and the stone or inner shell is the *endocarp*; only the kernel is the true seed.

QUESTIONS ON CHAPTER X.

How are leaf-buds protected from the weather?
What is the axilla? When can it be seen?
What is remarkable of the horse-chestnut leaf?
What are lymphatics, and secreting glands?
Give some common example of secreting glands.
What is said of the perfume of plants?
In which part of the plant is the seed perfected?
Give examples of a *raceme*, and of a *spike*.
Which are called the male organs of a plant?
What is required to the perfecting of the seed?
Name the various parts of the pistil.
What is meant by fruit in botany?
Distinguish the pericarp, sarcocarp, and endocarp?
What is the only true seed?

CHAPTER XI.

Vegetable Physiology.—We have thus traced the external form of the plant from its root to its ripened fruit. We now proceed to notice its internal structure and component parts, which are called the *elementary organs* of the plant. All the parts which we have mentioned, seem to be produced from the combination of the elementary organs, membrane and fibre. These are so fine as to require the use of powerful magnifiers to detect the peculiarities of their formation. There are three principal forms in which membrane and fibre are combined. In *cellular* tissue,

4—2

elementary membrane, or *woody fibre*, and *vascular tissue*.

On placing thin slips of pith, or what is better, a small portion of garden rhubarb, which has been slightly boiled, under a magnifier, it is found that it is composed of a number of small cells or bags placed side by side, and of very irregular formation.

Cellular Tissue.—These hollow bags are filled with fluid, and compose the cellular tissue which is found in every part of a plant. Besides the little sacs, there are to be seen very fine threads or filaments, and these tubes contain a spiral thread still finer; these are called vessels. The pith of the alder shows a similar arrangement of cells, but of an irregular form, which is caused by pressure of the cells one upon another. They are variously formed in other plants.

These cells thicken as they grow old, and cease to help in the nourishment of the plant; the cells are then empty, and only the cell-walls remain. The cell-walls are composed of carbon, oxygen, and hydrogen; whereas the cell-contents also contained nitrogen, in addition to the other three. The cells, while living, increase by dividing, each separate cell becoming two, in consequence of a division growing up its centre; so soon as this dividing tissue is completed, the cell separates into two.

Cell Contents.—These cells contain various substances in different plants; thus, in the potato the cells are filled with tiny granules of potato starch; in many seeds they are filled with oils, in the sugar-cane and beet-root they contain sugar. While the potato is good, the starch is kept in store, but as spring returns, the starch softens, and is used up or fed upon by the young plant when germinating

The fibrous tissue is composed of longer cells than the cellular tissue, generally much longer, and

BOTANY. 41

growing narrower at each end. As these grow old, their sides thicken and become hard, and form the wood of the plant.

The Stem and Sap.—The stems of plants are divided into two classes : those which grow externally, and are called *exogenous* or dicotyledous, such as are usually found in temperate regions, and those which grow internally, and are called *endogenous*, such as the date, and other palms, cocoa-nut, sugar-cane, and most other trees in tropical climates. They are also called *monocotyledons*.

Endogenous trees have no real bark, from the peculiarity of their growth. A hollow stem shoots up, and within this hollow, layer after layer of *inner bark* is deposited, until the trunk becomes a solid mass, and can grow no larger. The result is extreme hardness of the outer bark, a familiar example of which is the common cane.

Exogenous trees, like those which we see everywhere around us, are more complicated in form. While the wood grows from outside, the bark is endogenous, and grows from within; as the wood increases outwardly layer by layer, and year by year, the bark increases in a similar manner from within. As the inner coats of bark press upon the outer, it cracks and becomes rough, as we commonly see it in our bushes, shrubs, and trees.

The exogens are called dicotyledons, because they are all produced from a seed which has two cotyledons, or fleshy lobes, such as may be seen in the pea, bean, acorn, and other seeds.

The Pith.—If the stem of a dicotyledonous tree be cut across, it will be found to consist of the following parts. The *pith*, the *medullary* heath, the *wood* and the *bark*. In herbaceous plants, the *pith* occupies the greater part of the stem, also in some rapid

4—3

growers like the alder; but in old trees, its cells are thickened or filled up, so as to differ very little from the wood.

The Sheath.—The medullary sheath surrounds the pith, and is a tissue of spiral vessels and ducts, which carry water and other nutriment from the root to the leaves, with which it has a direct communication, and in which the food undergoes various changes, as noticed elsewhere.

Fig. 11. **The Wood.**—Next comes the wood, which is seen to be in rings in any plant which is more than two years old. By these rings the age of the tree may be generally known, as a new ring is added each year. It is worthy of notice that some tropical trees shed their leaves oftener than once a year, and form a ring each time, so that in such cases the age of the tree cannot be so easily ascertained.

Some sceptics, trusting to the number of rings, have tried to show that there are living specimens of tropical trees, which date back beyond the Deluge. These years should be reckoned as periods of vegetation, and not actual years, as would be the case in the exogens of temperate climates. The layers nearest to the centre are hardest, as the vessels are filled up with the secreted juice turned to solid wood. For this reason heart of oak, and other central wood, is used where great durability is required. The outer layers of wood are less hard, and are called *alburnum*, and it is through this part of the tree that the sap ascends from the root to the leaves.

The Bark.—The inner part of the bark is called *liber*, and is the source of new roots, buds, and leaves. Hence it is that old trees often flourish, after the ripe wood has been hollowed out. The *epidermis*, or outer bark of the tree, is continually decaying or sheathing off, after it has been cracked by the pressure of the *liber* within, and the outer surface of the liber as constantly supplies its place.

In the cork oak-tree of Spain and Portugal the epidermis is very thick, and it is this which furnishes the cork that is so valuable to us.

Monocotyledons (see Fig. 12) are easily distinguished from others; if one be cut across, a piece of cane, for example, it will be seen that they consist of bundles of woody fibres, and ducts or passages, dispersed among a mass of lengthened cellular tissue, and crowded together toward the outside. It increases by the formation of successive new bundles of tissue in the centre of the wood, and its diameter can never increase after the epidermis becomes hard, hence the crowded state of the outer ducts near the edge.

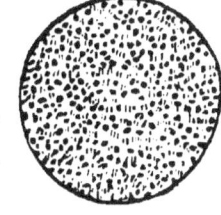

Fig. 12.

These little tubes give an example of capillary attraction, as spirits of wine and other volatile fluids will rise through them, if a short piece of cane be dipped into the fluid.

QUESTIONS ON CHAPTER XI.

What are the elements which compose plants?
Of what is cellular tissue composed?
In what plants may it be easily observed?
Of what are the cells formed? What do they contain?
How are the stems of plants divided?
What is an exogenous plant? Give examples.
By what other names are exogens known?
Name some examples of dicotyledons.
What is the medullary sheath? what its duty?
How may the age of an exogen be known?
Which is the hardest part of the tree, and why?
What is alburnum? What is liber?
What is the epidermis? What is cork?
How do you distinguish the wood of an endogen?
What example of capillary attraction is given?

CHAPTER XII.

SYSTEMATIC BOTANY, OR CLASSIFICATION OF PLANTS.

THE first attempt at a classification of plants was made in 1583 by Cæsalpinus, an Italian, who divided them into trees and herbs. The trees he arranged in two classes, based upon the situation of the ovary or seed-vessel, and the herbs into thirteen classes, also agreeing with their seeds and seed-vessels.

He was followed and improved upon by Robert Morrison, who was professor of botany at Oxford in 1660; but the most complete and philosophic method for a long time, was that of John Ray, the eminent English naturalist.

This method gave place to that of Linnæus in 1738, which was very generally adopted, and remained in vogue until Bernard de Jussieu, adopting the primary divisions of Ray, laid the foundation of the Natural System, which was perfected, or greatly improved by his nephew, Antoine L. de Jussieu, and which is still followed in France.

The system of Linnæus is very interesting. He divided plants into twenty-four classes, each consisting of two or more orders. The first eleven *classes* were founded on the number of the *stamens*, and their *orders* on the number of the pistils.

The twelfth contains plants with not more than twenty perigynous stamens, the thirteenth those with similar stamens, more than twenty in number. The fourteenth and fifteenth made their orders to depend on the form and structure of the *fruit ;* the sixteenth, seventeenth, and eighteenth, made the orders to depend on the number of the *anthers*, and so on to the twenty-fourth, which comprised all flowerless or cryptogamous plants.

THE NATURAL SYSTEM.

The Natural System, as somewhat modified by De Candolle, begins with the highest forms of plants, and descends to the lowest. Jussieu began in the opposite direction, from low to high.

The vegetable kingdom admits of two great divisions.

I. Flowering plants, called Phænogamæ, Cotyledoneæ, or Vasculares.

II. Flowerless plants, called Cryptogamæ, Acotyledoneæ, or Cellulares.

These are again divided into sub-divisions, classes, and orders.

Division I.—Flowering plants, or Phænogams, subdivided into *Exogens*, or Dicotyledons, *Endogens*, or Monocotyledons, and *Rhizogens*.

Exogens have embryo with two seed leaves, as pea or bean; stem increased by external layers, and leaves furnished with a network of veins. (See Fig. 13).

Endogens.—Embryo with one or many alternate seed leaves, stem increased by internal growth of cellular tissue and fibre, veins of the leaves running parallel, and flowers with three, six, or nine *petals*.

Fig. 13.

Exogens, sub-divided into four classes.

Class I. **Thalamifloræ**, from thalamus, the receptacle, including flowers that have calyx with separate divisions, and corolla with distinct petals. Petals and stamens inserted in the receptacle or thalamus. Example—The Ranunculus, or buttercup tribe, the poppy, cabbage, and many others.

Class II. **Calyciflorae**, having petals and stamens inserted in the calyx. Examples—the tea-tree, cashew-nut, pea, bean, the broom, and many other pod-bearers.

Class III. **Corolliflorae**, flowers having a corolla with the petals united, and not inserted in the calyx. Common examples — Azalea, rhododendron, heaths, &c.

Class IV. **Monochlamydeae**, or **Acorolliflorae**—flowers without petals, and having only a single floral envelope. Examples—Marvel of Peru, the cockscomb, love-lies-bleeding, beet-root, and many others.

The **Endogens**, which include one-fifth of known plants, are sub-divided into

Class I. **Dictyogenae**.—Veins of the leaves in form of a network; wood of stem in perennials arranged in a circle with a central pith. Examples—The yam, flowering rush, orchis, asparagus, palm, ginger, banana.

Class II. **Petaloideae**.—Leaves with parallel veins; flowers, a coloured floral envelope, or of whorled scales. Examples—Water plantain, lily family, hyacinth, tulip.

Class III. **Glumiferae**.—Flowers, glumaceous or scaly, consisting of imbricated scales, called bracts. Veins of the leaves parallel. Examples—The cotton sedge, and the grasses, including all the corn tribe, and the bamboo.

Rhizogens.—These form a connecting link between the flowering and flowerless plants, and have been placed differently by various botanists.

They have an embryo without seed-leaves, the fruit springing from a thallus. The stem is a mass of cellular tissue without fibre, and is furnished with cellular scales in place of leaves.

They are few in number, parasitical in growth, and, as in the case of the Rafflesia, appear to be a mere flower growing on the branch of a tree, having large fleshy lobes like the fungi. They have all a bitter flavour and strong smell, and are found only in tropical climates.

QUESTIONS ON CHAPTER XII.

Who first attempted to classify plants?
Name other botanists who succeeded him.
What system was long in use?
How did Linnæus arrange plants?
Who invented the Natural System?
Where does Candolle begin his classification?
What are the two great divisions?
How are the Phænogams sub-divided?
Name the four classes of Exogenous plants.
What proportion of the vegetable kingdom are Endogens?
Give some examples of Dictyogenæ.
What is peculiar of the Petaloideæ? Examples?
What is meant by glumaceous?
What are Rhizogens? What is the chief example?

CHAPTER XIII.

Division II.—**Cryptogamia,** or flowerless plants.

These are sub-divided into *Acrogens* and *Thallogens*.

Acrogens have the embryo without seed-leaves, and plants with a distinct stem bearing leaves or branches. They have no primary root, and only grow in one direction upwards.

This includes the fern tribe, the floating pepperwort, the lycopodium, or club-moss, and other mosses, and the hepaticæ or liverworts.

Thallogens represent the lowest form of vegetable life, sometimes consisting of a single cell, and in their highest development are only a collection of cells.

This sub-division includes the lichens, fungi, or mushroom, and toadstool tribe, the charas, a water-plant, and the seaweeds.

Cryptogamia, or flowerless plants.—These are ferns, mosses, lichens, fungi, and algæ, or seaweed, and are called flowerless because they have neither stamen nor pistil, nor any seed proper; but in the case of ferns are produced from or propagated by very small bodies called sporules, which in ferns are found under the frond or leaf. They are believed to be composed of cellular tissue without spiral vessels, but their composition is less understood than that of the flowering plants.

Mosses, which seem to be among the lowest of the kind, are found in damp places, on rocks, trees, walls, &c., chiefly in cold or temperate climates. The moss is reproduced from sporules which are found at the end of a stalk, in an urn or theca, which is closed with a lid called an operculum. There are 800 different species of mosses.

Lichens are stemless and leafless, consisting of a tough wrinkled substance called a *thallus*, and of various colours. These are reproduced from sporules, which are formed within the cellular tissue of which the plant is composed. This, like moss, grows in cold and damp climates, and there are said to be more than 2000 different species.

Fungi, such as the mushroom, toadstool, &c., are of various form and quality, some being excellent food, and others poisonous. Of rapid growth, they as quickly perish.

Algæ are water-plants growing either in fresh or

salt water. Of these seaweed affords numerous and curious examples, and this tribe approach nearest to the lowest grades of the animal kingdom, so that in some examples of both it is difficult to decide which is the animal and which the vegetable. Much seaweed was formerly burned for its ashes, called kelp, and is still used as manure.

QUESTIONS ON CHAPTER XIII.

What is the meaning of Cryptogamia?
What example is given of the Acrogens?
What represents the lowest order of plant life?
Name the various Thallogens.
The chief examples of Cryptogamous plants.
How are the Cryptogams propagated?
Describe especially the seed of the moss.
What is a lichen? How many mosses are known?
Where are these plants usually found?
What are the chief examples of Fungus?
What is said of their growth?
What are Algæ? What is said of this tribe?
What use is or has been made of seaweed?
How many varieties of lichen are known?

CHAPTER XIV.

THALAMIFLORÆ AND CALYCIFLORÆ.

Class **Thalamifloræ**.—This sub-class includes those exogenous plants which have both calyx and corolla, and have the *sepals* of the calyx, the *petals* of the corolla, and the stamens or seed-vessels, all growing separately and distinct from each other.

As in other classes,

Fig. 14.

there is a great diversity in the size of the members of this.

The best known examples are the common buttercup, poppy, water-lily, wallflower, violet, flax, lime-tree, and orange.

These are commonly arranged in eight groups called orders, and are named as follows:—

Order 1. **Ranunculaceæ**, including aconite, anemone, buttercup, clematis, Christmas rose, marigold, peony, larkspur, and columbine. These usually grow in damp places, and are all poisonous.

Order 2. **Papaveraceæ**, including all the poppy tribes. These are all possessed of narcotic or sleep-giving properties.

Order 3. **Nymphaceæ**, containing the water-lilies and lotus, all being water-plants.

Order 4. **Cruciferæ**.—This order is in the vegetable kingdom what the order Ruminantia is in the animal. It includes the cabbage, turnip, cress of all kinds, mustard, and others which are found in all climates, and are all health-giving, and therefore valuable to man. It may be known everywhere by its *four* sepals and petals in form of a cross.

Order 5. **Violaceæ**, including all the violet tribe.

Order 6. **Linaceæ**, of which the flax is the common example.

Order 7. **Tiliaceæ**, containing the lime-tree.

Order 8. **Aurantiaceæ**, containing the orange, and lemon, shaddock, grape fruit, and others, all natives of hot climates.

Class **Calyciflorae.**—This sub-class contains those plants which agree in having four or five sepals and petals, and numerous stamens, more or less fixed to the calyx.

The type of the order is the dog-rose, or common wild rose.

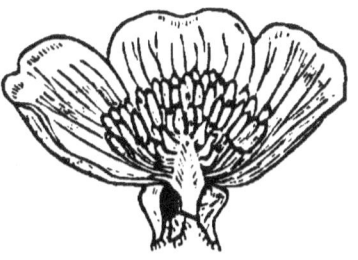

Fig. 15.

It is very extensive and important to man and other animals, as it contains valuable fruits and vegetables, as well as beautiful flowers. It is usually divided into *eight* orders.

Order 1. **Leguminosae**, which includes all those plants such as the pea, bean, gorse, and others, which have papilionaceous flowers, or flowers like the wings of *papilio*, a butterfly, adhering to the calyx.

Besides the common vegetables mentioned, it contains also the acacia, laburnum, ebony, and tamarind trees.

Order 2. **Rosaceae**, containing all the roses, the apple, pear, strawberry, and cinque-foil, and all the amygdalous or *almond-like* plants such as the almond, cherry, plum, apricot, peach, laurel, &c.

All the latter contain an active poison in their kernels or seeds, called Ferrocyanic, or Prussic acid.

Order 3. **Cucurbitaceae**, including the cucumber, (from *cucurbita*, a gourd,) melon, and the various gourds and vegetable marrows.

Order 4. **Grossulareae**, including all the gooseberry, currant, &c. These have usually green flowers, and frequently prickly stems.

Order 5. **Umbelliferae**, containing all the carrot and parsnip tribe, the hemlock, fennel, and parsley. This order is known by the growth of the flowers in *umbels* or bunches.

OUTLINES OF SCIENCE.

Order 6. **Lorantheæ**, including the misletoe and some other parasitic plants, usually without a corolla.

Order 7. **Crassulaceæ**, containing the houseleek, stone-crop, and others, all of which have succulent, or fleshy leaves.

Order 8. **Compositæ.**—The plants of this order have small flowers collected together into dense bunches or heads, surrounded by a case called an *involucre*.

The simplest example of this order is the dandelion; it also contains the sunflower, artichoke, groundsel, daisy, lettuce, endive, coltsfoot, chamomile, and others.

QUESTIONS ON CHAPTER XIV.

What plants are included in Thalamifloræ?
What is said about the size of the plants?
Into how many orders is this class arranged?
What is the best example of Ranunculaceæ?
Where do they grow, and what is their peculiarity?
What order contains the poppy plants?
For what are the poppy plants remarkable?
What are included in the Nymphaceæ?
Which is the most important order of the class?
What animal order does it resemble?
Give examples of the order Cruciferæ.
How may this order be easily known?
What plants belong to the fifth order?
To which order does the flax plant belong?
What is included in the Tiliaceæ?
What fruits are contained in Aurantiaceæ?
Where are the Aurantiæ found?
What is the type of class Calycifloræ?
Why is this class very important to man?
How is this class divided? Name them in order.
What are included in the Leguminosæ?
What are the chief examples of Rosaceæ?
What is peculiar to the kernels of amygdalous plants?
To what order does the cucumber belong?
What is notable in the flowers of Grossularcæ?
Why are the umbelliferous plants so called?

What plants belong to this order?
What order includes the misletoe?
What is peculiar to the Crassulaceæ?
Name the chief flowers of the order Compositæ.
What is peculiar to the flowers of this order?

CHAPTER XV.
COROLLIFLORÆ.

Class **Corolliflorae.**—This sub-class contains eight orders which may be known by the corolla or flower-cup being *monopetalous*, that is, instead of being divided like the dog-rose, or daisy, the corolla is in one piece, as the primrose.

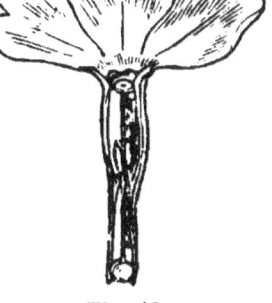

Fig. 16.

Order 1. **Boragineæ**, including the forget-me-not, hound's tongue, and the borages.

Order 2. **Primulaceæ**, including the common primrose, cowslip, oxlip, pimpernel, moneywort, &c.

Order 3. **Convolvulaceæ**, found twining in gardens and hedges, contains the various kinds of convolvulus and bindweed.

Order 4. **Oleaceæ**, the olive, lilac, ash, and common privet.

Order 5. **Gentineæ**, the gentian, yellow-wort, centaury, and buckbean, all valuable in medicine.

Order 6. **Solonaceæ**, or nightshades, includes several poisonous plants, such as the deadly nightshade, mandrake, stramonium, or thorn-apple, henbane, and tobacco. It also includes the potato and the egg-plant.

5—3

Order 7. **Labiatæ.**—This order contains the fragrant lavender, and the useful herbs sage, peppermint, thyme, horehound, the dead nettle, rosemary, and ground ivy.

Order 8. **Scrophulariaceæ**, including the digitalis or foxglove, the snapdragon, toad-flax, speedwell, broomrape, and euphrasia or eye-bright. The digitalis and euphrasia are valuable medicines.

EXOGENS.

Class **Monochlamydeæ.**—The plants of this subclass are distinguished by having only a single floral envelope, the flower being in one piece. It includes most of our timber and other large trees.

It is also divided into eight orders.

Order 1. **Polygonaceæ**, containing buckwheat, dock, sorrel, and knotgrass.

Order 2. **Urticaceæ** (from *urtica*, a nettle).—This order includes the common nettle, the hop, hemp, mulberry, and fig, the first three of which have very strong fibres.

Order 3. **Corylaceæ**, including the oak, beech, chestnut, hazel, and hornbeam.

Order 4. **Salicaceæ** (from *salix*, a willow,) contains all the willows, the sallow and osier, and the poplar and aspen trees.

Order 5. **Betulaceæ** (from *betula*, a birch,) contains the alder and birch trees. The alder makes the finest charcoal, and the bark of the birch is used in tanning.

Order 6. **Ulmaceæ**, the elm-tree tribe, very durable timber.

Order 7. **Pinaceæ**, or *coniferæ*, including the various pine and fir trees, and also the yew, cypress, and

BOTANY.

juniper. The leaves are all narrow and hard, and the wood is resinous.

Order 8. **Lauraceæ**, all aromatic plants, such as the camphor, laurel, bay, cinnamon, and sassafras.

QUESTIONS ON CHAPTER XV.

How are the Corollifloræ easily known?
Name the eight orders of this class.
What is meant by the word Monopetalous?
To which order does the forget-me-not belong?
What plants belong to the Primulæ and Gentineæ?
In what are the Gentineæ valuable?
Where are the Convolvulaceæ found?
Give examples of the order Oleaceæ.
To what order do the potato and tobacco belong?
What is peculiar to many of this order?
To which order do lavender and sage belong?
Name some valuable plants in Scrophulariaceæ.
How is the class Monochlamydeæ known?
Name its eight orders in succession.
What examples are given of Polygonaceæ?
Which order contains many fibrous plants?
Which order contains many aromatic plants?
What is peculiar to the leaves and wood of Coniferæ?
What order supplies the best charcoal?

CHAPTER XVI.
ENDOGENÆ AND ACROGENÆ.

Class **Endogenæ**.—This class includes seven orders, all of which have the roots growing from within the radicle of the seed, no central pith, and no outer bark. They have leaves with veins parallel to the mid-rib, or centre rib, (See Fig. 17,) sepals, petals, and stamens, in groups of three, six, nine, or twelve, and the seeds have only one seed-leaf.

Fig. 17.

56 OUTLINES OF SCIENCE.

Order 1. Palmaceæ, includes the cocoa-nut, flax, date, and other palm trees, all of which have leaves and flowers only at the top. The fruit is generally a nut, as the date or cocoa-nut.

Order 2. Liliaceæ.—This order consists chiefly of bulbous plants, as the various kinds, the tulip, onion, garlic, and asparagus.

Order 3. Orchidaceæ, containing the varieties of orchis, many of which have a great similarity in their flowers to insects, birds, or flies—as the bee-orchis, fly-orchis, and others of a similar kind.

Order 4. Iridaceæ, including the crocus and various flags.

Order 5. Juncaceæ, the soft rush, a common rush.

Order 6. Cyperaceæ, including the bull-rush and sedge.

Order 7. Graminaceæ (from *gramen*, grass,) contains all the grasses and varieties of grass-seed so important both to man and animals. The chief examples are the common wheat, oats, barley, rye, rice, seeds, meadow, and other grasses.

Class Acrogenæ.—Plants of this class have neither fruit nor flower, but commonly the spores or seed-vessels are arranged on the leaves. The root grows from any part of the seed, the thick parts are in a zigzag form; the leaves have either no veins, or simply forked veins.

This class is divided into seven orders.

Order 1. Filices.—This includes all the various ferns.

Order 2. Lycopodiaceæ, including the Lycopodium and club mosses.

BOTANY. 57

Order 3. **Equisetaceæ**, the horsetail which grows in marshes.

Order 4. **Musci**, the hair-moss, and those which grow on damp wood or stone.

Order 5. **Fungi**, the various mushrooms, toadstools, puff-balls, mildew, and smut. Many of the Fungi besides the mushroom are said to be good for food, but great care is required in the use of them.

Order 6. **Licheneæ**, including the yellow cup and other lichens.

Order 7. **Algæ**, the various kinds of sea-weed, of which there are an enormous number, nearly all adapted to grow under water. Some are used as articles of food, and all form good manure.

QUESTIONS ON CHAPTER XVI.

Name the peculiarities of the Endogens.
Name the orders of the class Endogenæ.
How do you distinguish an endogen leaf from an exogen?
In what order are the sepals, &c., usually placed?
What are the chief examples of the Palmaceæ?
What is peculiar to the growth of that order?
Which contains bulbous plants? Of what kind?
What is remarkable about the Orchidaceæ?
Give examples of orders four, five, and six.
Which is the most important of these orders?
What is peculiar to the class Acrogens?
Name its seven orders in succession.
Name the examples of the order Filiceæ?
Where do the Equisetaceæ grow? And the Musci?

CHEMISTRY.

INORGANIC AND ORGANIC.

CHAPTER XVII.

CHEMISTRY teaches us the nature of those elements that compose all the substances with which we are acquainted, the mode of their combinations, and the properties of the compounds which result from them.

This science is comparatively modern, but its results and discoveries have been as useful as they are wonderful. Every art and every branch of commerce has been indebted to it for development and improvement.

Alchemy.—The students of this science, previous to the seventeenth century, were chiefly those alchemists, as they were called, who had an opinion that the baser metals could be changed into gold by the aid of something which they called the "philosopher's stone." Many persons followed this pursuit in the middle ages, wasting their substance and their lives to no purpose, so far as gold was concerned, but their labours resulted in the discovery of many useful substances, and paved the way for a more intelligent study of the subject under present notice.

Chemistry is divided into *Inorganic* and *Organic*. The former treats of the simple bodies, or elementary substances, namely, those which cannot be further analysed or separated, while organic treats of their

CHEMISTRY.

various compounds, chiefly in the animal and vegetable kingdom.

Elements, or Simple Substances.—These, in ancient times, were said to be earth, air, fire, and water. The discovery of elements has greatly increased during the present century, and the chemist has shown that the original elements, as they were called, are all compounds, or, as in the case of fire, the product of combustion. Thus air is composed of oxygen, nitrogen, and a little carbonic acid; water, of oxygen and hydrogen; and earth of various substances, which may be analysed or loosed from one another, and shown to be elements.

Table of Elements.—The number of these at present is sixty-five; twenty years ago only fifty-four were enumerated—as chemists proceed in their researches, they will probably discover more. So long as a substance cannot be separated into any component parts, it is called an element, as gold, silver, and other metals. The following is the list of known elements, which are so called because they cannot be separated or further reduced:—

Aluminium	Chlorine	Lanthanum	Norium
Antimony	Chromium	Lead (Plum-	Osmium
Arsenic	Cobalt	bum)	Oxygen
Barium	Copper	Lithium	Palladium
Beryllum	Didymium	Magnesium	Phosphorus
Bismuth	Erbium	Manganese	Platinum
Boron	Fluorine	Mercury (Hy-	Potassium
Bromine	Gold (Aurum)	drargyrum)	Rhodium
Cadmium	Hydrogen	Molybdenum	Rubidium
Cæsium	Indium	Nickel	Ruthenium
Calcium	Iodine	Niobium	Selenium
Carbon	Iridium	Nitrogen, or	Silicon, or
Cerium	Iron (Ferrum)	Azote	Silicium

Silver, or	Sulphur	Thorium	Uranium
Argentum	Tantalum	Tin, or	Vanadium
Sodium, or	Tellurium	Stannum	Yttrium
Natrium	Terbium	Titanium	Zinc
Strontium	Thallium	Tungsten	Zirconium

Of the above, fifty-two are metallic elements, and thirteen non-metallic. Of the non-metallic, hydrogen, nitrogen, and oxygen are gases; chlorine and fluorine are vapours; bromine is a fluid; boron, carbon, iodine, phosphorus, selenium, and sulphur are solids. Those of them which form salts in combination with other substances, are called *halogens*.

Affinity, or Chemical Attraction.—Chemical affinity differs from the attraction of cohesion, and of gravitation, because it changes the nature of the substances upon which it acts. The results of affinity are as important as they are surprising. For example—two *gases* will form a *solid*. If a jar be held over hydrochloric acid, and another be held over a solution of ammonia, and the two jars be placed mouth to mouth, the two gases will combine and form the solid sal-ammoniac, which is seen in the form of a white smoke, and crystals are deposited.

Two *fluids* will form a *solid*. Add sulphuric acid to a solution of chloride of calcium, and a solid sulphate of lime will be formed.

Spontaneous Combustion is formed by throwing turpentine upon a mixture of nitric acid and strong sulphuric acid.

Also, if a few drops of sulphuric acid be dropped on a mixture of powdered loaf sugar and chlorate of potash, it will take fire spontaneously.

Combustion under warm water is caused by passing a stream of oxygen through a tube upon a piece of phosphorus.

CHEMISTRY. 61

Affinity changes colour, and is useful in testing various substances. If a solution of iodide of potassium be mixed with another of sugar of lead, both colourless, the result will be a *yellow* powder.

If iodide of potassium be mixed with a solution of corrosive sublimate (a preparation of mercury), a beautiful red powder is formed.

If a few drops of tincture of galls be dropped into a solution of sulphate of iron, a deep black colour will result, which is the base of the common black ink. Many similar interesting examples of change might be mentioned.

QUESTIONS ON CHAPTER XVII.

What does Chemistry teach?
What were the old chemists called?
How were the old alchemists employed?
Of what does Organic Chemistry treat?
What was the old notion about simple elements?
What are the component parts of air and water?
What is the number of the simple elements?
Why are they so called? Have they increased lately?
How many of them are metals? and non-metallic?
What is a halogen? What is affinity? How does it affect substances?
Give some examples from two gases; from two fluids.
What is said of spontaneous combustion?
How can we secure combustion under water?
How is affinity especially useful?
What colours result from compounds of iodine?
How is a deep black colour obtained?

CHAPTER XVIII.

Definite Proportion, or Combination.—When we speak of chemical compounds, we refer to the fact that the ingredients which form them are always united in them, in the same proportion as to weight.

In water, which is a compound of oxygen and hydrogen, the weight of the oxygen is *always* eight times that of the latter, and the same may be said of other compounds. Common salt contains $35\frac{1}{2}$ parts of chlorine to 23 of sodium, Glauber's salt always 80 of sulphuric acid to 62 of soda, and so on.

Various Compounds.—It is also remarkable that the same elements combined in different proportions produce a different result. This is especially noticeable in the compounds of oxygen and nitrogen. The simplest of these combinations is that of nit*rous* oxide, or laughing gas, consisting of 28 parts of nitrogen, and 16 of oxygen; the next nit*ric* oxide containing the same portion of nitrogen with twice the quantity (32) of oxygen; then follow nit*rous* acid, conbining 28 nitrogen and 48 oxygen; hyponi*tric* acid, which equals 28 nitrogen and 64 oxygen, and lastly, the very corrosive and powerful nit*ric* acid, which consists of 28 nitrogen and 80 oxygen.

Here it will be seen that the portion of nitrogen is always 28, while the proportions of oxygen increase from 16 to 32, to 64, to 80, or to twice, thrice, four, and five times the portion in nitrous oxide. The same law applies to other compounds, and there is no intermediate proportion. Should the element in a compound be more than 16, it is sure to be 32 or more, and if more, 48 or more, &c.

Laws of Combination.—The knowledge of these definite proportions is of the utmost value to those who have to manufacture chemical compounds.

They know that the same compound, whether obtained in Europe or America, consists invariably *of the same ingredients*, and that those elements are always contained *in the same proportion by weight*.

Chemical Symbols.—Every compound is represented by its equivalent symbols, and the proportions of each element contained in it. Thus water, which is always composed of *two* atoms of hydrogen and *one* of oxygen, is represented by the formula H_2O.

When a candle is burned, there is a consumption of two elements, carbon and hydrogen. The carbon burning in the air unites with two parts of oxygen, and forms carbonic acid, which is noted down as CO_2, while two parts of the hydrogen unite with one of the oxygen and form water, H_2O. The products of this combustion, therefore, are carbonic acid and water, and in a similar manner every known chemical compound is symbolised.

Chemical Language.—Many of the names of chemicals are derived from the alchemists, such as spirit of wine, spirit of salt, &c., and many of the older elements have two names, as sulphuric acid, and its common term, oil of vitriol, sulphate of copper, bluestone, &c.

Compounds are termed *primary* when they consist of only two elements, as water of oxygen and hydrogen (H O), sulphuric acid of sulphur and oxygen (SO_3), common salt, chlorine and sodium (NaCh), vermilion of mercury and sulphur (Hg_2S), &c.

Secondary, when formed of two *primary*, as sulphate of soda, which is composed of sulphuric acid and soda ($SO_3 + NaO$).

Ternary compounds are those which contain two secondary substances, or three elements, as common dry alum, which contains sulphate of soda, potash, oxygen, and aluminium.

Binary Compounds.—The chief of these are those of the non-metallic elements, such as oxygen, sulphur, and chlorine, with each other, or with the metals.

Compounds of oxygen are called *oxides*. Common rust is an oxide of iron; red lead, an oxide of lead.

Chlorides. — Compounds of chlorine are called *chlorides*, as common salt, the chloride of sodium, calomel, the chloride of mercury, &c.

Compounds of sulphur are called *sulphurets* or *sulphides;* vermilion is the sulphide of mercury, and galena the sulphide of lead.

Protoxides, &c.—The same elements as we have seen in the union of nitrogen and oxygen, form combinations of various proportions. The first is called a *protoxide*, from protos, first, as nitrous oxide. The next combination is called a *deutoxide*, as nitric oxide, and so on with the tritoxide, up to the hyper or peroxide, which is the highest combination of oxygen with any other element.

Bases.—In these combinations spoken of as *oxides*, the element which combines with the oxygen is called the base; thus, lead is the base of the oxide of lead, iron, the base of oxide of iron, and so on.

Acids.—Oxides which contain a large proportion of oxygen, will redden vegetable blues, and have an acid taste. They are therefore called *acids*, as sulphuric acid, nitric acid, and others.

Salts.—These are combinations of acids with oxides. Thus sulphuric acid combining with potash, forms the salt called sulphate of potash; when it combines with soda, sulphate of soda, or Glauber's salt; and when it combines with magnesia, it forms Epsom salt, or sulphate of magnesia. If these salts contain an acid ending in *ic*, as sulphur*ic*, they are called sulph*ates*, but if the acid contained ends in *ous*, they are called sulph*ites*.

The same rule applies to the other acid combinations, as nitrates, chlorates, hydrates, phosphates, &c.

QUESTIONS ON CHAPTER XVIII.

Give some examples of definite proportion.
Name some remarkable compounds of oxygen.
What are the proportions of nitrous and nitric acids?
Why is a knowledge of this subject important?
What are the chemical symbols for water?
What results from the burning of a candle?
What is a primary compound? a secondary?
What elements are found in common alum?
What are the compounds of oxygen called?
What is oxide of lead? What is a chloride?
What is a protoxide? What is a peroxide?
What is meant by the base of an oxide?
When much oxygen is contained, what is it called?
What is a salt? How is sulphate of potash formed?
Distinguish between sulph*ates* and sulph*ites*?

CHAPTER XIX.

THE GASES—OXYGEN, HYDROGEN, AND NITROGEN.

Oxygen.—Of these, the most abundant, and perhaps important, is oxygen, as it is more widely diffused than any other. It forms one-fifth of the volume of air, and eight-ninths of water by weight, besides entering largely into the composition of all the materials of the globe; neither animals nor vegetables can live without it.

Oxygen is colourless, and heavier than common air, is easily procured by heating chlorate of potash in a retort, and is used for many brilliant experiments. It will support combustion, but will not burn, so that if a piece of lighted charcoal or red-hot wire, or a small piece of phosphorus, be placed in a jar of it, those articles burn very brilliantly until the oxygen is exhausted, without fear of explosion. Its chief compounds, which are very abundant, are the oxides of the various metals.

Hydrogen.—This gas is colourless, and without taste or smell, and it forms one-ninth part by weight of all water. It is the lightest known gas, being fourteen times lighter than air, and sixteen times lighter than oxygen. It burns, but will not support combustion, so that by putting a light to the mouth of a jar of it, it is exploded at once with a pale yellow flame.

Lightness.—On account of its lightness, it is used for filling balloons, but as it does not exist in a separate form, it has to be prepared. The simplest mode of obtaining it, is to send a galvanic current through water, which separates the hydrogen from the oxygen.

The usual mode of preparation, is to pour a mixture of sulphuric acid and water on small pieces of zinc or iron filings, when the water is decomposed, and hydrogen immediately set free. If the mixture be placed in a bottle, and a tube be thrust through the cork, the gas may be burnt at the end of the tube. This is called the philosophic candle. Care must be taken to allow all the common air to escape before the light is applied, because a mixture of oxygen and hydrogen is very explosive, and both the cork and tube would probably be blown out of the bottle.

The Drummond Light.—This powerful light is formed by burning a double stream of oxygen and hydrogen on a piece of lime, whence it is called the oxyhydrogen or lime light. It may be seen to a distance of sixty miles, and is especially useful in lighthouses.

Nitrogen, or **Azote.**—This gas constitutes about four-fifths of the atmosphere, and is without colour, taste, or smell. It does not burn, nor will it support combustion, and when separated from oxygen will not support life. It is found in abundance in nitre, salt-

CHEMISTRY. 67

petre, and other nitrates, and in coal, and is a constant ingredient in plants and animals.

Compounds.—Several of the compounds of nitrogen are of great importance, the chief of which are nitric acid, which, as we have seen, is a compound of oxygen 28, and nitrogen 80, and which is so powerful that it dissolves many of the metals, and all animal matter. If a small piece of copper wire be placed in nitric acid, it begins instantly to dissolve, and the acid turns green, while reddish fumes escape. Nitrogen and hydrogen form the strong pungent compound ammonia, from which hartshorn is made. It is also a very important ingredient in food.

Chlorine and **Fluorine** are Gaseous Vapours.— Chlorine, which is of a yellowish green colour, abounds in nature. It is a chief part of common salt, which is chloride of sodium, as well as of rock salt, some earths and some mineral waters. It is easily obtained by heating a mixture of manganese and hydrochloric acid, in a retort, and may be detected by its pungent suffocating smell. As it destroys all vegetable colours, it is very important as a bleaching agent.

Chlorine has a strong affinity for metals, and powdered antimony, if shaken in a jar of chlorine, will burn brilliantly and form a chloride of antimony. A similar result follows the introduction of copper or gold-leaf. It does not unite with carbon, so that a lighted candle burns very badly in it, and the carbon falls to the bottom of the jar. If a paper soaked in turpentine (a compound of carbon and hydrogen) be introduced, it takes fire instantly, a" chlorine has a great affinity for hydrogen. The hydrogen unites with the chlorine, and the carbon is deposited as soot on the side of the jar. Besi*f* being useful in bleaching, it is valuable as a destroyer of noxious smells and gases.

Fluorine is much less common than chlorine, and may be obtained from fluor spar, when crushed and heated with hydrochloric acid. It has such a strong affinity for potash, a component of glass, that it cannot be put in a glass vessel, which it would dissolve by uniting with the potash in the glass.

It is used for writing or etching on glass, by simply drawing the pattern with the fluid. It will not combine with oxygen.

Bromine is so called from its disagreeable smell; it is found in mineral waters, and in all salt deposits, and is usually obtained from the *bittern* which is left in salt water, when the salt has been taken from it by evaporation. It is of a dark brown colour, destructive to the eyes, if allowed to touch them, and is a deadly poison. It is employed, to a small extent, in medicine and the arts.

QUESTIONS ON CHAPTER XIX.

Which gas is thought to be most abundant?
Of what does it form the chief proportions?
How is it procured? What are its properties?
Show that it supports combustion, but does not burn. Its chief compounds.
How is hydrogen obtained? Name its properties.
Describe what is called the Drummond light?
Of what gas is the air largely composed?
Name its chief properties. How is it obtained?
Name some of its chief compounds.
What are chlorine and fluorine?
How is chlorine obtained? What are its properties?
What happens when turpentine is introduced?
To what uses is chlorine applied?
Whence is fluorine obtained? How is it used?
What is the result of its affinity for potash?
What is the peculiarity of bromine?
How is it obtained? and how used?

CHAPTER XX.
NON-METALLIC SOLIDS.

Boron.—This is a dark olive substance found chiefly in volcanic districts in union with soda in the form of borax. It is of much use in the manufacture of glass and porcelain, in the manufacture of solder, and in medicine. Most of the metallic oxides, when heated with borax, form coloured glasses.

Carbon.—This element abounds in nature, in the diamond, in common coal, charcoal, blacklead, and other forms. The diamond is carbon in its purest form, but what is called blacklead is often pure carbon, or a compound of carbon and iron, called graphite. It never contains any *lead*. Lampblack and ivory-black are other forms of almost pure carbon.

Carbon combines with numerous other elements; its principal compounds are carbonic acid, carbonic oxide, and its various compounds with hydrogen.

Carbonic Acid, a compound of carbon and oxygen, is found everywhere in nature, and was called by the ancients spiritus lethalis, or deadly air, and afterwards fixed air. It is a colourless gas, with a sharp taste and pungent smell, and will neither burn nor support combustion, and is therefore destructive to human life. It is the *choke-damp* of miners, and often causes death to persons who go into old wells, lime-kilns, distillers' or brewers' vats.

Preparation.—Carbonic acid is prepared easily from limestone, marble, chalk, and other carbonates of lime, by pouring diluted sulphuric acid over the broken pieces of chalk. It may be also made, in its purest form, by burning charcoal in oxygen. As it is heavier than common air, it sinks to the bottom of the vessel in which it is contained, and may be poured like water from one vessel to another. A lighted candle is instantly extinguished by it, and it is therefore the custom, before going into a well or other place where its presence is suspected, to let down a light first. If there is life for the light, there is life for man, and it is safe.

Soda Water is a strong solution of carbonic acid gas in water, the gas having been forced into the water by a heavy pressure, hence the bubbles, caused by the gas escaping when the cork is withdrawn.

As we have shown in physiology, carbonic acid is given off from the lungs, and when we breathe into a glass of clear lime-water it becomes clouded, as a carbonate of lime is formed by the union of the carbonic acid with the lime.

Carbonic Acid in water dissolves the carbonate of lime in rocks, and forms *stalactites* and *stalagmites* in limestone caverns. It is prevented from accumulating to a dangerous extent in the atmosphere, by the absorbing action of the vegetable kingdom, which feeds on it to a large extent. By this means, also, plants are supplied with it that grow in soils which do not contain it, such as siliceous, or flinty soils.

Carburetted Hydrogen.—The compounds of carbon with hydrogen are both interesting and useful. The most common is coal-gas, which depends for its lighting power on compounds of carbon and hydrogen, called heavy carburetted hydrogen, or olefiant gas, which burns with a bright red flame, and light carburetted hydrogen.

Firedamp.—Light carburetted hydrogen, is that gas which is generated in stagnant pools and in marshy places where vegetables are decomposed under water, and in some coal-pits, and burns with a light blue flame. It is called Will-o'-the-Wisp, and is very dangerous in coal mines, when mixed with common air, as it will explode with great violence. It is called *firedamp* by the miners. As the carburetted hydrogens are colourless, invisible, and without smell, the danger is not suspected until the miner comes in contact with this enemy.

The Safety Lamp.—To preserve them from its evil influence, the safety lamp was invented by Sir Humphrey Davy.

The principle of it depends on the fact that flame, which is gas at a white heat, will not pass through small tubes or openings, such as the meshes of wire gauze. As the flame of the safety lamp is surrounded by wire gauze, the presence of firedamp is detected by its influence on the flame within, and the miner is warned of danger.

QUESTIONS ON CHAPTER XX.

What is Boron? How is it usually found?
Mention its various uses.
What are the various forms of carbon?
What is blacklead? What is lampblack?
What are the compounds of carbon?
What are the properties of carbonic acid?
How is it prepared? What are its uses?
What are stalactites and stalagmites?
What prevents the accumulation of carbonic acid?
What is a good test for carbonic acid?
What are the useful compounds of carbon?
What are its dangerous compounds?
What is heavy carburetted hydrogen?
When does carburetted hydrogen become explosive?
What is the principle of the safety lamp?
How is it useful to the miner?
Distinguish between firedamp and chokedamp.

CHAPTER XXI.

Iodine is found chiefly in sea-weeds, and in sea-air, and is obtained by burning sea-weed in masses. It is a bluish-black crystalline solid, which crumbles readily when pressed. When heated it gives a beautiful violet colour, and when placed in contact with a piece of dry phosphorus, affords a good example of spontaneous combustion, as the two burst into flame and form the iodide of phosphorus.

A solution of iodine of potassium is detected instantly; if starch be added, it turns a deep blue; if nitrate of silver, a pale yellow; if acetate of lead, a bright yellow; if corrosive sublimate, a fine scarlet; while perchloride of platinum turns it to the colour of claret or port wine.

Phosphorus.—This substance is found in abundance in union with lime, magnesia, guano, and bones of all kinds. The bone derives its strength from the phosphates of lime and magnesia, which compose its hard material, and from which the phosphorus of medicine is prepared. This element is a component of grain of all kinds, and of peas and beans, and much of the various phosphates is used by agriculturists to enrich the land. Phosphorus is kept in water, as it melts at a temperature of 115°, and would disappear in the form of vapour, if left uncovered.

Very much of this substance is used in making lucifer matches, as the heat caused by gentle friction will cause the compound to take fire.

Selenium is a rare substance much resembling sulphur, in connection with which it is occasionally found, and at present of no use either in medicines or the arts.

Sulphur is a common element found free in abundance, in volcanic districts, and is extensively diffused in combination with other substances, and with the metals. The best is brought from Sicily. It is of a lemon colour, and is highly inflammable, burning with a blue flame and forming the pungent sulphurous acid, the smell of which is so well-known, and so disagreeable. This acid is useful for bleaching purposes, and is used to bleach wool, silk, sponge, and straw plait.

Sulphuric Acid.—Sulphur burned in oxygen produces the powerful fluid called sulphuric acid, which is highly useful both in medicine and the arts under the name of oil of vitriol, and which, when heated, dissolves most of the metals. The sulphur burned in oxygen produces sulphurous acid, but the use of a small portion of nitre in the combustion, changes it into sulphuric acid, or oil of vitriol; the vapour passes out of the chamber in which it is produced, and which is lined with lead, through a passage into another chamber, the floor of which is covered with water, in which the acid is condensed and collected.

Sulphuric acid is used in numerous chemical manufactures, as well as by dyers, calico printers, and bleachers.

Refiners of metal, soap and candle makers, soda-water manufacturers, and numerous other trades, are greatly dependent on its effects.

Affinity for Water.—Vessels containing this acid must be kept covered up, as it has a powerful attraction for water, and will grow weaker by absorbing aqueous vapours from the atmosphere.

Great care is required in mixing it with water, as they rapidly combine and evolve great heat, and often break the vessel in which they are mixed. Two parts of acid suddenly mixed with one part of water, will cause heat equal to boiling.

Experiments.—Sugar turned to charcoal. If you place some powdered loaf sugar in sulphuric acid, or make a strong syrup, and pour diluted acid into it, the acid takes the water from the sugar, and leaves nothing but black charcoal behind, showing clearly that the chief components of loaf sugar are *water* and *charcoal*.

To show its bleaching powers, burn a little under a glass in which a rose is suspended. The sulphurous acid will quickly destroy the colour of the flower.

QUESTIONS ON CHAPTER XXI.

Where is Iodine found? What are its peculiarities?
What example does it give of spontaneous combustion?
What colour does it give with corrosive sublimate?
Of what is phosphorus a component?
What is said of its temperature? How is it kept?
What is selenium? Where is the best sulphur found?
Mention some of its properties? Its chief compound?
In what art is sulphur especially useful?
What persons are dependent on sulphuric acid?
For what has it a great affinity? How is it kept?
What follows the sudden mixture of vitriol and water?
What are the chief components of sugar?
How is it proved? Is sulphur a good conductor?
Show how to test its power of bleaching.

CHAPTER XXII.
THE METALS.

Aluminium.—This is a metal of a light grey colour, which is obtained from alumina, or pure clay. Clay is usually termed an earth, but it is really a rust or an oxide of the metal aluminium, that is, an earth which has been produced by burning that metal in oxygen. It was discovered by Davy in 1808. As carbon is the essence of the diamond, so is alumina of the ruby and sapphire—one also forms the basis of coal and coke, the other of pipeclay and earthenware.

Antimony is a bluish-white brittle metal, found in

combination with sulphur, which passes away in vapour at a white heat, and takes fire spontaneously in chlorine.

Arsenic, the common poison, is a grayish-white metal more than five times heavier than water, which is found in combination with the ores of cobalt and nickel, from which it is separated. It oxidates rapidly on exposure to the air, forming a white powder. It burns spontaneously with a blue flame in chlorine.

As it has often been the cause of death by poison, both by accident and otherwise, it is well to know that the best antidotes for it are calcined magnesia, or the hydrated oxide of iron.

Tests.—Its presence may be detected in any liquid by the addition of a little sulphuric acid, and placing thereon a piece of zinc : if the liquid contains arsenic, a gas will be evolved called arseniuretted hydrogen, which, if set on fire, will deposit tiny globules of metallic arsenic on the side of the tube, or on a piece of glass held over the flame.

Barium is a greyish heavy metal which is obtained from barytes.

Bismuth is a white metal, nearly ten times the weight of water. Its chloride is used by artists as Spanish white or pearl white.

Cadmium is a metal similar to tin, found rarely in combination with the ores of zinc.

Calcium is the base of lime, or calx, which is only an oxide of this metal, as alumina, or pure clay, is an oxide of aluminum. It was discovered by Davy in 1808, and has a brilliant lustre like silver. When heated it burns into lime. It is the base of marble, limestone, chalk, alabaster, and all the other carbonates of lime, which abound in nature ; it is found in the teeth and bones of animals, and is of the utmost importance to builders and cement makers.

Plaster of Paris.—Calcium is also the chief component of gypsum, or plaster of Paris, which has a strong affinity for water, and forms with it a hard cement.

The lime is procured by burning the chalk or carbonate of lime with coals in a lime-kiln, when the heat drives off all the carbonic acid. The purest lime is obtained from marble, which is carbonate of lime, that has been exposed to great volcanic heat.

Cerium is obtained from cerite, which is a very rare mineral, white and brittle; it will dissolve in hydrochloric acid, but not in sulphuric acid.

Chromium is a greyish-white metal, which is found in combination with the ores of lead, copper, and especially chrome iron ore. It forms with oxygen the green oxide, and in various compounds is used in great quantities by dyers, artists, and bleachers.

Compounds.—Chromic acid, with the salts of lead, gives a *yellow* compound called chrome yellow; with black oxide of mercury, an *orange;* from the bichromate of potassium when heated, beautiful *green* crystals are obtained. The red tint of the ruby is said to be caused by the presence of chromic acid. The chromate of mercury is a red colour.

Cobalt is a metal of a reddish-grey colour, which melts readily, and is much used in the colouring of glass and porcelain. Chloride of cobalt is used as a "sympathetic ink," which will turn blue when heated.

Copper.—This well-known metal, which is called brass in scripture, is found in all countries, and is the only one of a red colour. It is of great use, on account of the readiness with which it unites to form various compounds, such as *spelter*, a hard solder formed with equal parts of copper and zinc; *brass*, which contains 16 parts of copper to 9 of zinc; *bell-metal*, 3 parts copper, and 1 of tin; *bronze*, 9 parts copper to 1 of tin. Copper is also used in making *pinchbeck*, 5 parts copper and 1 zinc; *Manheim* gold, 3 parts copper, 1 zinc, and a little tin.

In making sovereigns and other gold coin for circulation, nearly one-twentieth copper is mixed with the gold to harden it.

From its slight tendency to rust, copper is used to cover or sheath the hulls of ships, and from its pliability and strength it is extensively used as wire in the vastly-increasing telegraph apparatus.

Solutions of Copper.—In many places copper is held in solution in streams of water, and as it has a strong tendency to unite with iron, old iron vessels are placed in the water, which become coated with it, and by this means large masses of copper are collected.

Malachite, the splendid green mineral found in Russia, is a carbonate of copper, which is found in a native state, and is quarried out like marble.

Uses and Tests.—Sulphate of copper and its other salts have either a green or a blue tinge, and are much used in the arts and in medicine. Many experiments may be tried with solutions of copper, and as they are more or less poisonous, and are, notwithstanding, used to colour pickles, it is desirable to know how to detect its presence. A solution of ammonia gives a deep *blue ;* hydro-sulphuric acid, a deep *black ;* prussiate of potash, a *reddish-brown ;* and if a clean knife or plate of steel be placed in a solution containing much copper, it will be at once precipitated on the steel.

Sources of Copper.—The copper used in England

is mostly obtained from copper pyrites in Cornwall, but it is sent across the Bristol Channel to be smelted or separated from its ore. This metal is found in a native state in North America, and is abundant in Australia.

Other Metals.—*Didymum, erbium, glucinum, lanthanium, lithium, norium, terbium, thorium, yttrium,* and *zirconium,* are very rare metals, having qualities and forming compounds similar to those of aluminium, but not of sufficient importance to be noticed in our limited outline of chemistry.

QUESTIONS ON CHAPTER XXII.

How many metallic elements are known?
What is aluminium? What is clay?
Of what jewels is alumina the essence?
Of what manufacture is it the base?
What is antimony? How does it act with chlorine?
What is arsenic? Where is it generally found?
How may the presence of arsenic be detected?
What are barium, bismuth, and cadmium?
What is calcium? Where is it found?
Of what substances is it the base?
What is lime? How is it made?
What is cerium? What are the properties of chromium?
By whom is chromium found especially useful?
What colours are made from its compounds?
What is cobalt? In what art is it much used?
Where is copper found? Name some of its compounds.
For what is copper much used? Why?
What compound of copper is found in Russia?
Name some tests for copper in solution.
Where is the English copper found? Where smelted?
Where found in a native state?
Name any other rarer metals in this chapter.

CHAPTER XXIII.

Gold (*aurum*) is found in most countries, but only in the metallic state, and generally in grains, or nuggets in quartz, or in small particles in the sands of rivers. It is separated from the sand or from the quartz when crushed, by washing.

Carat Gold.—Pure gold is of a reddish-yellow colour. The gold of our coinage is mixed with copper, and is called 22 carat gold, because it contains twenty-two twenty-fourth parts of the pure metal. The gold of jewellers contains eighteen of these twenty-fourths, and is therefore called 18 carat gold. It is nearly twenty times heavier than water.

Electrotyping.—The peroxide of gold is much used in electrotyping; that salt of gold being obtained by dissolving gold in *aqua regia, i.e*, a mixture of nitric and hydrochloric acids. The value of gold is increased by its wonderful malleability, or capability of being beaten out.

Iron (*ferrum*).—This valuable metal is found in nearly all parts of the world, either native in small quantities, as magnetic iron ore, as hæmatite or red oxide of iron, or in its most common and abundant form, as clay ironstone, or some carbonate.

Pig Iron.—Magnetic ore and hæmatite are easily melted, but the ironstones require an intense heat, and the addition of lime or limestone. The lime combines with the clay and sets the metal free; it sinks to the bottom of the tall furnaces, and is let out boiling into beds of sand, and thus forms pig-iron, or iron in its first stage of manufacture.

Malleable, or Pure Iron.—In this state it is an impure and brittle cast-iron. and if it be required to separate the impurities from it, and to render it tough to bear heavy strains, it has to be heated over again once or more, and placed under heavy steam-hammers and rollers, which squeeze out the carbon, and it is afterwards cut into bars for sale to iron-founders.

Steel.—Cast-iron is converted into steel by heating it for forty or fifty hours in contact with charcoal. The iron absorbs the carbon, and the bars are then heated and welded or hammered together, which increases their tenacity or toughness. Sometimes the process is repeated, and the steel acquires its property of extreme hardness.

Silver Steel.—By adding one 500th part of its weight of silver, the hardness of steel is greatly increased; but its hardness depends also on the way in which it is cooled after being greatly heated. If it is cooled slowly, it becomes softer, and it is often heated several times, and cooled quickly or slowly as it is required to be hard or soft. This is called *tempering* the steel.

Sword Blades.—The famous Damascus and Indian sword blades, which have been struck against iron and stone, without breaking, are made by welding pieces of iron and steel together. They are now successfully imitated by Mr. Wilkinson, of London, whose process of manufacture may be seen in the Museum of Geology, with numerous other chemical and mineralogical curiosities.

Rust.—Iron does not oxidise readily in a dry atmosphere, but does so quickly if allowed to get wet. The oxide is what is called rust, and its influence may be observed at the bottom of old iron railings which have not been kept well-painted.

All the crystals and salts of iron are of a green

colour, and are of much use in the arts, as well as in medicine.

Dyers and ink-makers are indebted to salts of iron for their black colours, and glass bottle makers for their colouring matter.

Lead is a soft, flexible, and non-elastic metal, which is found in combination with other metals, or in a sulphuret called galena, from which it is separated by heat. It melts at the low temperature of 612°, oxidises quickly on exposure to the air, but the outer coat of rust preserves what is underneath it.

Type used in printing is formed of three parts lead and one of antimony. Pewter is a compound of lead and tin, and solder also in various proportions, according as it is required to be hard or soft.

Rust of Lead.—Lead oxidises rapidly when heated, and a grey film is formed on it, which may be seen in any plumber's ladle, and various compounds of lead are used in the arts and in medicine. Painters are greatly indebted to this metal for the *white lead*, which they use in great quantities, and which is obtained by exposing sheets of lead to the fumes of a vegetable acid such as vinegar.

Magnesium is a white metal of which the well-known magnesia is the only oxide, and is now extensively used to burn in gas where a brilliant white light is needed. It exists abundantly in nature as carbonate of magnesia, and as magnesian limestone. Epsom salts is a sulphate of magnesia.

Manganese is a greyish-white granular metal, which becomes quickly oxidised by exposure to air, on account of its affinity for oxygen. It is usually found mixed with iron, or some other metal. Its compound, called the black oxide of manganese, is much used in glass-making, and in the preparation of oxygen, and in the arts generally.

QUESTIONS ON CHAPTER XXIII.

How and where is gold found?
What is eighteen carat gold?
How is gold dissolved? How then used?
In what conditions is iron found?
How is iron-stone melted? Explain the process.
What is pig-iron? How is it made tough?
What is steel? Explain the process.
How is silver steel manufactured?
Upon what does its hardness greatly depend?
How are the Damascus sword blades made?
What causes iron to rust readily?
Of what colour are the salts of iron?
In what manufactures are they used?
What is lead? What is galena?
Name some properties of lead.
Of what is printing type composed?
What is white lead? How is it used?
What is magnesium? How is it used?
How does the oxide of magnesium exist in nature?
What is manganese? How is it usually found?
How is the oxide of manganese utilised?

CHAPTER XXIV.

Mercury, or Quicksilver, is a brilliant metal more than thirteen times heavier than water, and which has been known from very ancient times. It is generally found in the sulphuret of mercury, which is called *cinnabar*, but sometimes in combination with gold and silver, and is separated by roasting the ore with lime.

It is different from all other known metals, in being always in a *liquid* state at an ordinary temperature of the air, and only becomes solid when exposed to intense cold at 30° below zero.

Compounds.—These are numerous and highly useful

in medicine, as well as the arts. It is a dangerous metal, as its fumes are injurious to health, and a gentle heat will evaporate it. It easily unites with other metals, and is used in combination with tin to silver looking-glasses. It is also an essential article in the refining of silver. Its compound, calomel, is the sub-chloride of mercury, an active and powerful medicine, and the perchloride called corrosive sublimate is a deadly poison. Mercury is used in the manufacture of colours for artists and painters. Vermilion is a compound which is prepared from corrosive sublimate.

Tests for Mercury.—A solution of its salt gives with ammonia a *black* precipitate; with iodide of potassium, a greenish *yellow;* with chromate of potash, a *red*.

As accidents happen at times with corrosive sublimate, it is well to know that *the white of egg* is an antidote to it, if swallowed quickly after the poison.

Nickel is a metal similar in character and uses to cobalt. It is found abundantly in Germany, where the miners call it "nickel kupfer," or false copper. It is white like silver, and is extensively used in the arts. Nickel and brass combined, form German silver.

Platinum.—This is the heaviest of metals, its specific gravity being 21·5. It is found in South America, Russia, and Ceylon, but at present is only known to exist in small quantities. It will melt under the oxyhydrogen blowpipe, but not with any ordinary heat, and is therefore valuable to form crucibles or vessels in which other metals can be heated.

It does not oxidise or tarnish in either dry or wet atmospheres, and is not injured by the ordinary acids. It is five times the price of silver, but is much used in consequence of the characteristics mentioned, and especially in the manufacture of oil of vitriol.

Potassium.—This was discovered in caustic potash by Sir H. Davy, by means of the galvanic battery. It may be procured by subjecting cream of tartar to a great heat. It is of a bluish-white colour, soft like wax, and very brittle. When heated it gives off a green vapour. It has so great an affinity for oxygen, that it must be kept in naphtha in a stoppered bottle.

It floats on water, and takes fire uniting with the oxygen, and disappearing with a slight explosion. On this account it makes a very pleasing experiment.

Compounds of potassium, namely, potash in various forms, are of great importance in the arts. The principal are the carbonate, sulphate, and chlorate. The last is a dangerous compound, as it is very liable to explode. It is used in making lucifer matches, and in various detonating mixtures, and has caused the loss of many lives by accident.

A solution of potash is the base of the various kinds of soap, and it is also used in the manufacture of glass.

Sodium.—This metal is very much like potassium, and was discovered about the same time and in the same manner, by Sir H. Davy. It is light, and has a strong affinity for oxygen; it will also take fire, if placed on water, but does not burn so rapidly as potassium. It is equally important in art and manufactures, and is obtained in great abundance from common salt, which is a chloride of sodium.

Silicon (Si).—This metal was discovered in silex, or flint, by Sir H. Davy, and is a dull brown powder, with very little of the metallic in its appearance. Silex, or silicic acid (SiO_3), is abundant in the stems and outer parts of all the grasses, in the form of silicate of potash, and is the basis of the quartz family.

Porcelain.—Silex is a very important component with alumina in the manufacture of porcelain. The two earths being worked up together with more or less lime and potash, form a plastic clay, which, when shaped and partly baked, is painted with mineral colours which can stand a great heat. This manufacture has long been known to the Chinese and the people of Japan.

Glass.—This is also a combination of silicates, the finest flints being used for the fine glass, called flint glass on that account. Potash, or soda, causes the fusion or melting of the flint or sand, which is more commonly used, and, like porcelain, it is coloured with mineral oxides; *blue*, with oxide of cobalt; *green*, with protoxide of copper; *red*, by peroxide of iron; *purple*, by oxide of gold, &c.

Oxide of Gold is also used to gild glass, being afterwards burnished.

QUESTIONS ON CHAPTER XXIV.

What is mercury? Its other name?
How is it generally found? Its weight?
At what temperature does it freeze?
Name some of its compounds.
In what arts and manufactures is it used?
With what does mercury give a black precipitate?
Why is this metal dangerous? What is the antidote?
What is nickel? Where is it found?
What is its colour, and how is it used?
Which is the heaviest of metals? Where is it found?
What heat is needed to melt it? Does it rust?
For what purposes is it used? Its price?
Who discovered potassium? By what means?
Describe the properties and affinities of potassium.
What is said of chlorate of potash?
What metal is like potassium in nature and uses?

Whence is sodium obtained? Who discovered it?
What is silicon? Of what is it the base?
How is silex employed? Whence obtained?
Mention the colours that can be made from it.

CHAPTER XXV.

Silver, (Ag.)—This valuable metal has been long known. It is usually found in combination with other metals, and nearly always with *lead*. It is separated from the ore by bringing it into contact with mercury. The two metals form an amalgam from which the mercury can be expelled by heat.

Silver is very soft, and requires to be hardened by 18 parts out of 240 of its weight of copper, which is the standard silver of England. It dissolves easily in nitric acid, forming nitrate of silver ($NO_5 + AgO$), an extremely corrosive substance, which destroys animal tissues, and is much used in surgery. It is also the chief component of good marking ink.

All the salts of silver in contact with animal or vegetable matter, will turn black when exposed to light. They are much used in photography.

Tin (*Stannum*), in ancient times called Jupiter. This metal attracted the enterprising Phœnicians and Carthaginians to the British islands. It is seldom found alone, but usually in combination with ores of copper or zinc. It abounds most in granite veins, when it is called mine-tin. At other times it is found in alluvial deposits in small particles which are called tin-stone, and are of excessive hardness.

Uses of Tin.—This valuable metal has long been known to commerce, and was used as a compound with copper for the making of swords, knives, &c., until the method of working iron was discovered.

This metal tarnishes quickly, but oxidises or rusts slowly under moisture, and is therefore valuable for protecting other metals, as it is so malleable as to be reduced to extreme thinness, as in *tin-foil*.

Compounds of tin are also important in glass and porcelain making, and in dying, as it fixes certain colours on woven fabrics. The Romans bought it to mix with other metal to form their bronze helmets, &c., and numerous old furnaces remain, which prove that in ancient times it was extensively obtained. It was chiefly worked by Jews in the time of the Plantagenets.

Zinc (Zn.) is found abundantly in France and Germany as a zinc blende, a sulphuret, or in calamine, a carbonate of zinc. It is a bluish metal, hard and brittle, which rapidly oxidises when heated in the open air. Its compounds with copper are noticed in the article on that metal.

The chloride of zinc, *i.e.*, zinc dissolved in hydrochloric acid, is extensively used as a disinfectant, as it stays the decay of animal matter, and decomposes offensive gases. It is also used to saturate wood, as a protective against the dry rot. Zinc is now used extensively by builders and others, as it does not rust deeply, the oxide which rapidly forms, becoming a protection to the metal beneath.

QUESTIONS ON CHAPTER XXV.

How is silver usually found?
How is silver separated from its ore?
What is called standard silver in England?
In what is it easily dissolved? What is that called?
Describe some uses of nitrate of silver.
In what are compounds of silver much used?
What is said respecting their change of colour?
Where is tin found? What was it once called?
Name some of its properties and uses.

88 OUTLINES OF SCIENCE.

How does minc-tin differ from tin-stone?
Why is it especially valuable to coat other metals?
For what purpose did the Romans seek it?
Where is zinc found? In what forms?
What is chloride of zinc? What is its use?
Name some of its compounds with copper.
Why is zinc especially useful to builders?
Which are the heaviest of the metals?
Which are most malleable?
Which are most useful in the arts?
What metals oxide or rust easily?
Which of them resist oxygen best?

CHAPTER XXVI.
ORGANIC CHEMISTRY.

Organic Chemistry treats of those bodies, animal and vegetable, which have organs of increase, repair, and reproduction. Inorganic matter is the nutriment of plants, organic matter that of animals; grasses and other plants feed on the silex, potash, phosphates, &c., found in the soil, and animals feed on them, so that all living creatures may be said to feed on grass.

Organised substances contain but few elements; vegetables consist chiefly of carbon, oxygen, and hydrogen; animals of carbon, oxygen, hydrogen, and *nitrogen.* There are some exceptions, as in beetroot, sugar, maple, and unripe fruits, which contain nitrogen.

Plant Food.—The organic matter of the soil, which must consist of decayed vegetable matter, is continually forming carbonic acid, which is absorbed with water by the rootlets of the plant (See Botany). Hence the richness of those virgin soils of the tropics, on which the vegetable matter has been decaying century after century. Light, acting on the carbonic acid in plants, decomposes it, and sets free the oxygen. The tropical regions which are so rich in plants, furnish oxygen for the more temperate regions, and to those districts which, destitute of vegetation, are deficient in that element.

The principal elements of organic chemistry, which we can notice, are the various vegetable acids and their compounds, with the constituents of oils and fats.

Oxalic Acid, Ō or C_2O_3.—This acid is found in the common sorrel and many other plants, and also in union with lime in the bodies of animals. It is obtained by heating *one* part potato starch or sugar with *five* parts nitric acid.

When evaporated, the acid is seen in the form of needle-like crystals. Its solution is a violent poison, for which either chalk or magnesia are antidotes. An oxalate of potash, called salt of lemon, is used to remove iron-mould and other metallic stains from linen.

Cyanogen, Cy or C_2N.—This is a colourless gas, which is easily procured from the kernels of almonds, peaches, and other stone fruits, and from bay leaves. It burns with a rose-coloured flame. With hydrogen it forms the powerful liquid called hydrocyanic or *prussic* acid, which is so very destructive of life when swallowed or inhaled.

The combinations of cyanogen are numerous and important in art and manufactures. It is one of the chief elements of success in photography.

Urea, $C_2N_2H_4O_2$.—This element may be obtained from urine, of which it is an important part, or from the union of cyanic acid with dry ammonia.

Uric Acid, $C_{10}H_4O\,N_4$. — This acid is found in human urine and in that of all carnivorous animals, birds, reptiles, and many insects. The urine of birds and serpents consists almost entirely of urate of ammonia, and so also does guano, which is the decomposed excrement of sea-birds.

This acid has neither colour, taste, nor smell; it is thought to be the basis of the stones or calculi, that are occasionally found in the human bladder.

Acetic Acid, $C_4H_3O_3$, or $\overline{A}$; common name, vinegar— This acid is found in all liquids capable of the vinous fermentation; the alcohol absorbs oxygen and becomes acid, which it would not do unless it came in contact with yeast.

Much vinegar is made by mixing poor wine that has become sour, with a little vinegar, and exposing it to the air in casks containing the husks of grapes. Free access of air shortens the process, as the alcohol is more quickly oxidated.

Pyroligneous Acid.—One kind of vinegar is obtained by the destructive distillation of wood, which, though less agreeable to the taste, is cheaper, and is capable of being used to preserve meat, to which it gives a smoked flavour.

Creosote, $C_{14}H_2O_2$.—This is another product of the distillation of wood, which may be obtained from raw pyroligneous acid. It is an oily colourless liquid, with a pungent acid taste, and an odour of smoke. It coagulates albumen, and thereby prevents the putrefaction of animal matters.

Formic Acid, C_2HO_3.—This acid, so called from *formica*, an ant, was formerly obtained from those insects, but may now be had from a mixture of starch or sugar, with oxide of manganese, water, and vitriol. It is a clear, colourless, and caustic or burning liquid, with a strong smell, the vapour of which burns with a blue flame.

CHEMISTRY. 91

Tartaric Acid, $C_8H_4O_{10}$ or Co.—This acid abounds in the juice of the grape, tamarind, and other fruits, and is often found combined with potash on the inside of barrels in which French and German wines have been kept. It is readily obtained by adding one part *lime* to four parts of cream of tartar, and is very useful in medicine.

Citric Acid, $C_{12}H_5O_{11}$.—This acid is so similar to tartaric acid that they were thought to be identical until 1784. It is found in abundance in lemon juice, and is of great use in medicine, especially as a preventive of scurvy among seamen. It is also largely employed in the manufacture of lemonade, and by calico printers; it will remove iron-mould from linen, by forming a soluble compound with oxide of iron.

Malic Acid, $C_8H_4O_8$ or $\overline{M}$.—This acid abounds in the juice of apples, to which it gives the sour taste, and is therefore called malic, from *malus*, an apple-tree. It is also found in other fruits, and in the berry of the mountain ash, from which it may be easily obtained, by saturating their juice with lime.

Tannic Acid, or Tannin, $C_{18}H_5O_9$ or $\overline{Qt}$. — This acid was formerly called the astringent principle. It is obtained from the bark of the oak, horse chesnut, and other trees, and especially from the gall-nut of the oak-tree. In union with salts of iron it gives a deep colour almost black, and this, suspended in a solution of gum, is the common writing ink.

But it is of especial service in the manufacture of leather. After the hair has been removed by the action of lime, the hide is steeped for some days in a tank or pit containing bark and water, until the skin becomes changed into leather by the action of the tanning.

Sometimes animal bodies have been found whole and tanned in peat bogs, the flesh being preserved from putrefaction by the antiseptic quality of the peat juice.

Gallic Acid, C_7HO_3 or $\bar{G}$.—This acid is found free only in the mango fruit, but may be obtained also from the infusion of galls, oak bark, walnut, and other shells. Like tannic acid, it gives a black precipitate with salts of iron.

It will thus be seen that these various and often exceedingly different compounds are formed from the same elements, carbon, oxygen, hydrogen, and nitrogen, in different proportions.

QUESTIONS ON CHAPTER XXVI.

Of what does organic chemistry treat?
Distinguish between the nutriment of plants and animals.
What plants and fruits *do* contain nitrogen?
What is formed by decaying vegetable matter?
How is the carbonic acid set free?
What are the principal elements of organic chemistry?
Where is oxalic acid found? Its symbol?
What form do the crystals take? What are its properties?
What is cyanogen? How is it obtained?
Name some of its compounds and uses.
Where are urea and uric acid found?
In what substance does it abound?
What is acetic acid? How is it formed?
What is pyroligneous acid? For what is it used?
What is creosote? Name some of its properties.
From what was formic acid obtained?
How is tartaric acid obtained? Where often found?
What acid was thought to be the same thing?
How is citric acid employed? Whence obtained?
What other acid is found in many fruits?
What was once called the Astringent principle?
Name some uses of tannic acid.
Where does it exist in nature?
Where is gallic acid found? How is it used?

CHAPTER XXVII.
THE VARIOUS OILS AND FATS.

Fixed Oils.—*Oils* are either fixed or volatile. The latter can be distilled, but the fixed oils must first be decomposed. They are found chiefly in the cellular tissue of animals, or in the seeds and capsules of plants. Many nuts contain about 50 per cent. of oil, linseed 20 to 25 per cent., and hempseed 25 per cent.

All the fixed oils, while fluid, will leave permanent grease stains on paper. There are some which dry and form a hard coating when exposed to the air, such as linseed oil, nut-oil, and poppy-seed oil; but olive, almond, and rapeseed oil do not dry when so exposed.

Spontaneous Combustion.—Drying oils, when exposed on a large surface, such as shavings or cotton waste, absorb oxygen so rapidly as often to take fire, and produce spontaneous combustion; on this account great care is required in their use, among any similar light material.

The fixed oils are soluble in alcohol and turpentine, but not in water.

Glycerine (*glukus*, sweet) $C_6H_7O_5 + HO$ or $GlyO + HO$.—This is the sweet principle of oils and fats, and is a colourless syrup, which is obtained by heating equal parts of olive or other oil and fine litharge with a little water, and stirring it until a thick paste is formed. Glycerine is the basis of stearine, margarine, oleine, and other fatty compounds.

Stearic Acid, CHO, or St.—This fatty acid is obtained from animal fat, and purest from that of mutton. It is of great use in the manufacture of soap and candles. Stearate of soda is the basis of all hard soaps, and stearate of potash the basis of soft soap. The acid is used largely in the production of a stearine candle, nearly as good as wax.

Margaric Acid is very much like stearic, and oleic also, which is the basis of oleine, the chief constituent of the fixed oils.

Oleine is very useful to oil clocks and watches, because it does not congeal, except at a very low temperature. Steam and other engines are greased with more solid substances, on account of the great heat to which they are exposed.

Butter contains stearine, margarine, and oleine, and three other components, butyrine, caproine, and caprine, which may all be obtained by saponifying the butter, *i.e.*, mixing it with some alkali, and afterwards decomposing it with tartaric acid.

Spermaceti is found dissolved in spermaceti oil, in the skulls of various kinds of whales. It separates from the sperm oil after death, and is a valuable component of candles.

Soap.—This is a salt of the fat acids. It has long been known as in use among the barbarian Gauls, by whom it was used to render their hair shining, as is stated by Pliny.

Hard soaps are made with the fats and fat oils mixed with either soda or potash. Soda gives a harder soap than potash. If oil be used the soap is much softer, so that all the soft soaps are made with the drying oils.

Soap of potash may be made harder by the addition of common salt, as the chlorine of the salt unites with some of the potash in the soap, and forms chloride of potassium.

The detergent or cleansing properties of the soap depend on the fact that the neutral stearates, margarates, and oleates of potash and soda are decomposed by cold and hot water, and an alkali set free, which cleanses the clothes or flesh by forming a soap, with the fatty or dirty matters which soil them.

Mottled Soap, if genuine, contains salt of copper or of iron, by which it is made harder, and less water than the common kinds.

Hard Water is made softer by adding carbonate of potash or soda; where much lime is present the soap is decomposed and wasted, and tends rather to cling to the soiled clothes, than to wash or dissolve the dirt out of them.

Caoutchouc, also called india-rubber, is obtained from the milky juice of several plants which grow in hot countries. A mould of clay is covered with successive coats of the juice, which flows from the tree, and hardens outside the mould. It is of a yellowish colour when fresh, but is darkened by the air.

Beside the common use of caoutchouc as a rubber, it is much used, when dissolved in turpentine, for rendering various fabrics waterproof. It melts at 240°, and at 600° is dissolved into a vapour, which is again condensed into **caoutchisine**, the lightest liquid known. This article is used to dissolve oil-colours, and in making varnish, as it dries very rapidly.

Contraction.—Heat *contracts* it, while it expands almost every other substance; while being kneaded or punched to render it soft, great heat is evolved; this may be seen by simply stretching a piece of it, when heat is at once evolved.

Vulcanite.—Heated with sulphur to 300°, it combines with it, and forms the "vulcanised india-rubber" of commerce, which is remarkable for its permanent elasticity.

Caoutchouc burns with a bright flame, and in Cayenne and other places, where it is common, it is used instead of candles.

Gutta Percha is the juice of a tree which grows in Borneo, and other East Indian islands, and is obtained, like caoutchouc, by cutting a notch in the tree, from which the juice flows. At a temperature of 212° it becomes soft and plastic, so that it may be used for making mouldings, picture frames, and many other useful articles. Its chief value to the chemist is that it is impervious to water, and resists acids and alkalies, and it is therefore made use of to hold those articles.

It is a powerful insulator, or non-conductor of electricity, and is used in great quantities for covering telegraph wires, which, without some such protection, would part with their electricity as they passed under the earth's surface.

QUESTIONS ON CHAPTER XXVII.

What is the difference between fixed and volatile oils?
Where are fixed oils found? Which will dry?
How is spontaneous combustion sometimes caused?
In what are the fixed oils soluble?
What is the sweet principle of oil called?
Of what is glycerine the base?
What acid is especially needed for soap and candles?
What other fatty acids are used?
What is used for greasing machinery?
What substance contains all these acids?
What is spermaceti? Whence obtained?
How did the Gauls use soap?
What effect have soda and potash on soaps?
What is the peculiarity of mottled soap?
How may hard water be made softer?
What is caoutchouc? Whence obtained?
What are the chief uses of india-rubber?
What is vulcanised india-rubber? Its use?
Where is gutta percha found? What is it?
At what temperature do caoutchouc and gutta percha melt?
Of what special use is it to chemists?
Why is it much used in telegraphy?

CHAPTER XXVIII.

Starch, $C_{12}H_9O_9 + 2HO$.—This compound is found in abundance in the cellular tissue of wheat, potatoes, arrow-root, and many other plants, in the form of small white grains of various sizes and shapes. It is easily obtained by scraping the potato into cold water, then pouring off the milky liquor, we find starch among the settlement.

When a potato is frozen, the cells which contain the starch are burst by the expansion of the water, and the root becomes so unwholesome, as to be unfit for man or beast.

To obtain starch from wheaten flour, the flour must be fermented to sourness, that the gluten may separate from the starch.

Gum, $C_{12}H_{11}O_{11}$.—This well-known substance is obtained from the juices of many plants, as gum arabic, &c., soluble in water, and it forms mucilage, which may be precipitated from the solution by alcohol, and which forms with nitric acid what is called mucic acid. It has not the property of vinous fermentation.

Sugar, $C_{12}H_9O_9 + 2A_9$.—This name includes all substances which are capable of vinous fermentation, that is, which are changed by decomposition into alcohol and carbonic acid.

Cane Sugar is found in the juice of the sugar cane, beet-root, carrot, turnip, potato, and in the nectaries of many flowers. The juices are clarified with lime, and slowly evaporated at a low temperature, when crystals are formed of all the crystallizable matter, and the rest is run off as molasses or treacle.

Fermentation.—This is of several kinds. When starch is changed into sugar it is called *saccharine*, when sugar is changed into alcohol it is called *vinous*, and when alcohol is changed into acetic acid, or vinegar, it is called *acetous*. When malt liquor is fermented without yeast, a mucilage or gum is produced, and it is called the *viscous* fermentation. When a body undergoes decomposition, with disagreeable odours, it is called the *putrefactive* fermentation.

Alcohol, $C_4H_6O_2$.—This is a very strong volatile spirit, which is obtained in abundance by distillation from brandy, whiskey, and strong wines, but is also found in all the substances that contain starch or sugar. That which is made from damaged grain, contains an impure fatty matter, and is very injurious to those who drink it. It has a strong affinity for water, and is never found free from it, except when alcohol has been poured upon quicklime, when the lime will absorb the water contained in the alcohol, and the absolute alcohol will pass off as a vapour.

Ether is a very volatile fluid, produced by distilling alcohol with an acid. The most common is sulphuric ether, which is prepared by heating alcohol with sulphuric acid. It is much used in medicine, and has a powerful tendency to remove spasmodic affections.

Lignine, $C_{12}H_8O_9$.—This substance forms the tubes and cells of vegetable tissues, and is formed from starch by removing three atoms of its water. It may be easily obtained from sawdust. Flax, when dressed, consists almost entirely of lignine.

Gun Cotton.— This is another remarkable compound, obtained in the first instance from starch heated in strong nitric acid. It is now made by immersing clean cotton for about fifteen minutes in a mixture of equal parts nitric and sulphuric acids, keeping the cotton entirely covered during the process. The acid is then pressed out, the cotton washed with water, until it no longer reddens litmus paper. It is then dried in a current of air at about 212° of temperature.

It explodes by percussion or striking, and not by friction, and takes fire at 356° with little smoke. Its force is *four times* that of gunpowder, and it is much employed in blasting rocks, and in mining operations.

QUESTIONS ON CHAPTER XXVIII.

Of what is starch composed? Its symbol?
How may starch be obtained in a natural state?
What is mucilage? From what obtained?
What substances are classed among sugars?
What substances are formed by vinous fermentation?
Whence may cane sugar be obtained?
Describe the process of sugar making.
Name the various kinds of fermentation.
What is putrefactive fermentation?
How and whence is alcohol obtained?
What is contained in alcohol made of bad grain?
For what has alcohol a great affinity?
How can it be separated from water?
What is produced by distilling alcohol with acids?
What substance forms vegetable tissue?
How can lignine be obtained?
How is gun cotton made?
What will cause it to explode?
What is its force? How is it employed?

ELECTRICITY.

CHAPTER XXIX.

Electricity is the science which treats of that great force in nature, the electric fluid, about which comparatively little is known.

From very ancient times philosophers noticed the existence of some attractive power in amber. The Greek philosopher, Thales, of Miletus, called this subtle, unknown spirit, electricity, from *elektron* the Greek word for amber.

Pliny and others wrote of the attractive properties of amber, but do not refer to any further development of the subject, and it was only in the sixteenth century that Dr. William Gilbert made an attempt to classify some electrical phenomena.

His experiments led others in the seventeenth century to continue the study, and Dr. Wall, Newton, Boyle, and many others at home and abroad, down to our own Faraday, have increased the number of facts and experiments in relation to it.

Electricity is defined as a fluid which seems to pervade all substances, and which, when undisturbed, remains in a state of equilibrium. In this respect it is similar to heat.

Vitreous and Resinous Electricity.—If a glass tube be rubbed in a flannel, or in a soft dry silk handkerchief, it will attract small balls of dried pith, or pieces of tissue paper. The paper will hang for a moment upon the glass, and then fall off.

If a light downy feather be suspended by a thread of *white silk*, and the excited glass be applied to it, the phenomenon will be similar, the feather will cling for a few seconds to the glass tube, and then fall away. If the tube be excited afresh, and again applied to the feather, it will no longer attract it, but will *repel* it. The feather will move *from* the glass. But if a stick of sealing-wax be rubbed and applied to the feather, it will *attract* it. Similarly if the feather be attracted first by sealing-wax, and then repelled by the second rubbing, it will be attracted by presenting a glass tube. On this account it is supposed that there are two kinds of electricity—*vitreous* and *resinous*.

The Electroscope.—This is an apparatus more or less simple, by which we detect the presence of the electric fluid. It may be made by hanging balls of pith on silk threads, which when applied to the conductor, or any excited electric, will repel each other. The most delicate form is Bennett's gold-leaf electrometer, which has two leaves of gold inside a glass jar, attached to a brass knob, and by which the presence of the least quantity may be detected.

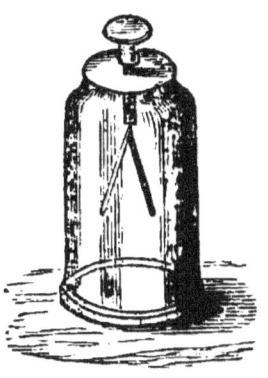

Fig. 13.

The Heat of Hair.—If a quantity of hair be woven and placed on the head of a wooden doll, and the doll fixed on the conductor of a machine, as the cylinder is slowly turned, the hairs on the head of the doll will stand out apart from each other, until something is placed near them to take a portion of their electricity.

If a boy be placed on a stool with glass legs, and holds a chain attached to the conductor, as the machine is worked, he will become similarly charged with electricity, his hair will stand out, and sparks may be taken from him, while he will feel as though cobwebs were being spread over his face.

Positive and Negative.—Faraday found that when silk is rubbed with *glass* it becomes negative, when rubbed with sealing-wax it becomes positive, and that in a similar way, in every case of rubbing or excitation, the rubber and the thing rubbed become the one *positive* and the other *negative*. On this account it is usual to speak of the vitreous as positive, and the resinous as negative. Bodies positively electrified *always* repel each other.

Electrics and Non-electrics.—There are some substances, such as those already mentioned, by which electricity can be easily developed; there are others in which it is very difficult, if not impossible, to do so.

The latter are called *conductors*, or non-electrics, and consist of the various metals, their ores and salts, charcoal, all fluids except oil, all saline and earthy substances, smoke and steam.

The principal *electrics* or *non-conductors* are gutta percha, glass, amber, sulphur, all resinous substances, wax, silk and cotton, feathers, wool, hair, paper, loaf sugar, dry air, oils and metallic oxides, hard stones, and dry earth. The one kind are *conductors*, because electricity passes through them, more or less easily; the other *non-conductors*, because they do not transmit it.

Various forms of Electricity.—The phenomena of electricity may be produced in several ways, and are therefore known by different names, as *Frictional*, which is produced by rubbing various electrics; *Magnetic*, which is seen in magnetism; *Thermo-electricity*, which is caused by heat; and *Galvanic* Electricity, which is the result of chemical action on certain metals.

Machines. — It is evident that only small quantities of the fluid can be obtained by rubbing glass or sealing-wax with the hand. It is therefore necessary to excite larger surfaces. This is done by the electric machine, which may be of two forms, either a glass cylinder, as in fig. 19, or a circular glass plate. This glass cylinder is so fixed as to press against a rubber, by which means frictional electricity is developed, over a large area. The fluid must be drawn from the earth, and therefore a chain is hung from the rubber to the ground, otherwise the rubber, which is insulated, that is, fixed on a glass support, would not be able to supply the fluid when excited.

Fig. 19.

The rubber is always negative when excited, and the cylinder positive. A piece of silk passes over the top of the cylinder to the conductor, to help retain the fluid, and the rubber is assisted in its work by an amalgam, formed of tin and quicksilver, which is smeared over it when the friction commences.

Leyden Jar.—It is desirable for various purposes to collect the fluid in quantities, and this is done by a Leyden Jar (fig. 20), or by a battery, when a powerful shock is required. A glass jar is coated with tinfoil, outside and inside, about two-thirds of the way up. Through the cork or wooden stopper a wire or thin metal rod is passed, to which a

Fig. 20.

small chain is attached, that touches the bottom. When the machine is turned, the electric fluid will pass into the jar, the inside will be *positively*, and the outside *negatively* electrified, or the inside will have more, and the outside less, than what is called its natural share.

Discharger.—Any machine which connects the outside with the inside, will restore the equilibrium, and for this purpose an instrument is used called a discharger, (fig. 21), with a glass handle. When one knob A is brought near the knob B of the leyden jar, the fluid passes from the inside through the metal conductor to the outer coating of the jar. This fact was first noticed by Cuneus, a Dutch philosopher, who was holding a glass of water, which held some of the electric fluid, and who, when he put his finger into the water, received a slight shock. By using a discharger with a glass handle, the equilibrium may be restored without any shock; but if one hand be placed on the outside, and the knob B be touched with the other, a violent shock is felt as the fluid passes through the arms and upper part of the body to complete the circle.

Fig. 21.

QUESTIONS ON CHAPTER XXIX.

In what substance was electricity first noticed?
By whom was it first named? Why?
When did Dr. Gilbert classify the phenomena?
In what respect is electricity like heat?
Name the kinds of electricity.
How is the vitreous fluid excited? And the resinous?
What is the electroscope? The most delicae form?
What happens to a head of hair electrified?t
How are the vitreous and resinous states named?
What is an electric? A non-electric?
Name the best electrics. The best conductors.

Name the various forms of electrical excitement.
How is the fluid collected in quantities?
Whence must it be drawn? By what means?
Describe the electric machine and leyden jar.
By whom was the discharging rod invented?

CHAPTER XXX.

ELECTRICITY—*continued.*

The Electric Shock.—It is a source of great amusement, if instead of a discharger, a number of persons are arranged so as to complete the circuit, one end holding the outside of the jar, and the person at the other end touching the knob at a given signal. The fluid in that case passes through the bodies of all who form the chain, giving a shock of greater or less severity, according to the size of the jar, or the quantity of fluid contained.

The Battery.—For experiments in which great power is required, what is called a battery is used. This is composed of several Leyden jars, which are connected both within and without, so that they can all be charged or discharged at the same time, and by this means thin wire is heated and even melted, gunpowder and spirits of wine set on fire, and many other similar experiments.

Darkness and Warmth.—By performing electrical experiments in the dark, the various phenomena are seen in greater perfection. As the machine is turned, the fluid is seen to pass round the cylinder to the conductor, and from the conductor to the jar or battery in a stream of bluish fire.

It is also to be noticed that a warm and dry atmosphere is always desirable for the success of electrical experiments, and also that the apparatus be quite free from dust as the fluid passes away from the surface quickly, if it be dusty.

Thunder and Lightning.—In hot weather the air becomes overcharged in parts with the electric fluids. Some clouds contain more than their natural share, or are positively electrified. When clouds thus charged come in contact with others containing less than their natural share, the electric fluid passes from one to the other with a flash and an explosion. The flash causes lightning, and the explosion causes the thunder. Many persons dread the thunder, but it is the lightning which is most to be feared, as it often does damage to trees and tall buildings which are not properly protected. We see the lightning before we hear the thunder, because light travels more rapidly than sound. We may tell the distance of such an electric explosion by noticing how many seconds of time pass between the lightning and its thunder.

Sound travels at the rate of 1140 feet per second: if we notice, for example, that 12 seconds elapse in the interval between the lightning and the thunder, we may say that the storm is distant from us about 12 times 1140 feet, or 13,680 feet, which being divided by 5280 feet in a mile, gives a distance of about $2\frac{3}{5}$ of a mile.

Lightning Conductors.—It has frequently happened during thunder storms that the electric fluid has struck tall buildings, such as factory chimneys, church steeples, towers, and trees, and ships. On this account it is dangerous to be under tall trees during a thunder storm.

Dr. Franklin, of America, proved that it could be attracted to the earth by sending up a kite, fastening a key to the lower end of the kite-string, and putting some silk (a non-conductor) between his hand and the key. The fluid was attracted down the string by the key, from which Franklin took sparks in abundance.

He suggested as a means of preventing accident, that a continuous metal rod should be carried up by the side of high buildings, to be pointed at each end; the upper end to be tipped with copper, or some other metal which does not rust, because oxides or rusts are bad conductors; the lower end to be buried in the earth or in a water-tank. By this arrangement, when the fluid passes near the metal point, it is attracted to it, and passes silently down into the earth.

Blasting Rocks.—If a strong charge of electricity be sent through thin wire, it will melt it; platinum wire is made red-hot by it. The knowledge of this fact is made use of by mining and other engineers to set fire to gunpowder. The powder has a wire passed through it, which is connected at either end with an electrical apparatus. On a given signal the circuit is completed, and the platinum wire inserted in the powder becoming red-hot, the powder is exploded, without danger to those who are performing the experiment. It may be sent any distance, if the conducting wire be properly insulated, as it may be, by coating it with gutta percha.

QUESTIONS ON CHAPTER XXX.

Describe what is called an electric shock.
What is the cause of the shock?
What is an electric battery? Its uses?
How is the brilliancy of experiments increased?
What atmosphere is most favourable?
What natural phenomena are caused by the electric fluid?
Why is lightning to be feared rather than thunder?
How may its distance be discovered?
At what rate do sound and light travel?
Has lightning ever been brought to earth?
State by whom and in what manner?
What is the effect of lightning on tall objects?
How can such accidents be prevented?
How is electricity made useful?
What wire is used? Describe the process.

CHAPTER XXXI.

GALVANISM.

GALVANISM and Electricity, if not the same thing produced by different means, are certainly intimately connected. Electricity is always produced by friction of some kind, while Galvanism is the result of chemical action only.

If a half-crown or a silver plate be placed *upon* the tongue, and a similar plate of zinc *under* the tongue, on bringing the metals into contact, there is a disagreeable acid taste produced. This is galvanism in its simplest form.

How Produced.—If plates of copper and zinc be soldered together in such order that each plate of zinc is connected with each plate of copper, and the connected plates be placed in vessels containing dilute acid, there will be an instant development of chemical action, and galvanic electricity is the result.

It is to be noticed that the latter is much less powerful in its action than frictional electricity, so that to produce any notable effects, batteries of considerable size, and containing numerous pairs of plates, are necessary.

When the zinc plate is placed in the battery, the acid does not act perceptibly upon it, until a plate of copper or platinum be also introduced, and the two brought into contact, which may be done by connecting them with a wire outside the battery. The circuit being completed, the zinc is at once oxidated by the acid, and becomes *negative*, while the other plate is *positive*, and the galvanic fluid flows from the positive to the negative.

Size of Plates.—It is remarkable that the intensity of the action does not depend much on the size of the plates, but on the number of the circuits. A tiny battery may be made in a thimble with the two metals and a little salt and water, but when great energy is required, a number of cells must be placed side by side, and connected by slips or wires of copper or any other good dry conductor.

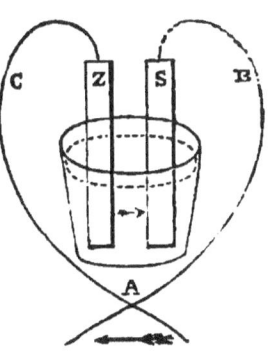

Fig. 22.

Galvani.—The philosopher, Galvani, who was the discoverer of the science, was professor of Anatomy at Bologna, about 1790, and he who first noticed the fluid did so, as we are accustomed to say, by accident. He was performing some experiments on frictional electricity, when he or his wife noticed that a small quantity of the electric fluid sent through the nerves of a frog, produced a violent agitation or convulsion of the muscles. He tried various experiments on animals, both living and dead, some of which were very cruel, but which plainly proved that the passage of the fluid through the nerves caused a shock to the nervous system.

The Leech.—A simple example of its effect may be seen by placing a leech on a piece of copper, and making it also touch a piece of silver. So often as it touches both metals, it receives a shock, and a continuation of such shocks will kill the leech.

Voltaic Piles.—Other experiments were tried by Volta, who was a professor at Pavia, in Italy, and from whom galvanism is sometimes called voltaic electricity.

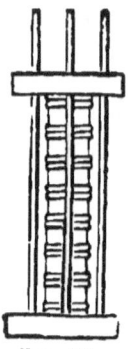

Fig. 23.

He formed what is called a voltaic pile, by placing discs of silver and of zinc in a pile, with cards soaked in dilute acid placed between each, as in fig. 23, where the lines represent plates of silver and zinc, and the card or cloth soaked in salt water, or dilute acid, and placed between them. By touching the lowest plate and the uppermost at the same time, a shock was received, which was repeated every time the *circle* was thus completed.

Advantage of Galvanism.—The electric fluid produced by galvanic action is of most use, because, though its action is less intense, it is continuous and apparently inexhaustible, while that developed by frictional electricity, however intense it may be, is discharged instantaneously. For example, if it be required to decompose a drop of water into its original elements, oxygen and hydrogen, it will require a very powerful shock of frictional electricity to perform it, but by means of a suitable instrument, the positive and negative wires of a galvanic battery may be introduced into water, and so long as the circuit remains completed, the water will be decomposed, the hydrogen will quickly ascend from the negative pole to the upper end of the tube, while the oxygen will collect in bubbles on the inside of it. The knowledge of this fact has enabled experimenters to provide those gases at a cheap rate, and to store them in gasholders, until they require to use them for experiments, or to give intensity to the light required in the exhibition of dissolving views, or of the oxyhydrogen microscope. The action is quickened by adding a little sulphuric acid to the water.

The Reduction of Metals.—If the tube be filled with a solution of acetate of lead, the gas will not be perceived to pass from the negative pole, but small metallic spikes like needles are formed on the negative wires. If the battery be powerful, and the tube contains a strong solution, as the process continues the lead is reduced or separated, and a small fern-like or tree-like mass of metallic spikes is the result.

Electrotypes.—The fact just noted, *i.e.* that the hydrogen from the negative pole separates metals from their oxides, has led to the process which is called electrotyping. A solution of the oxide of the required metal has to be placed on the vessel, and the object which is required to be coated with the metal is placed in contact with the negative pole; so soon as the circuit is formed, the metal begins to be deposited on the surface of the articles to be coated. In this way spoons, forks, dishes, &c., are plated with silver, while any object, such as medals, coins, casts, may be copied, as the finest lines are reproduced upon the deposited metal.

Galvanized Iron and Tin.—If sheets of iron or iron wire be placed in a solution of sulphate of zinc, and exposed to galvanic action, the zinc will be deposited on the iron, and what is called *galvanized* iron is produced. This is very useful, as being less liable to oxidation or rust than the unprotected iron.

The iron may be *tinned* in a similar manner, by placing it in a solution of oxide of tin.

QUESTIONS ON CHAPTER XXXI.

Describe how galvanism and electricity are produced.
What simple experiment developes galvanism?
How is it produced on a larger scale?
On what principle is the battery formed?
Of the metals employed, which becomes positive?
What is said about the intensity of the action?
How was the influence of galvanism discovered?
Show how a leech can be affected by it.

What is the principle of the voltaic pile?
What are the peculiar advantages of galvanism?
What are its effects on water?
How may the action be increased?
If a metal be contained in the solution, what occurs?
To what process has this led? Describe it.
How is iron galvanized?

CHAPTER XXXII.
MAGNETISM.

MAGNETISM is a peculiar property which exists in an oxide of iron, called *loadstone*, to attract iron and steel, and which property it communicates freely to iron in any form by contact or friction. It was first discovered at Magnesia, in Asia Minor, and has henceforth been called magnetes. It is now found in Norway and Sweden, in Elba, and in some parts of Asia.

Directive Power.—It is of especial use to man for what is called its *directive* power, which has an invariable tendency to arrange itself in a certain position with respect to the poles of the earth.

If a small magnet, or a steel needle rubbed with a magnet, be suspended or allowed to turn on a pivot freely, it will *always* point north and south. The knowledge of this fact has been made use of by mariners since the middle of the thirteenth century, and ships are now steered across the pathless ocean by the aid of the mariner's compass. Since that time nearly all the important discoveries have been made by navigators, who without the compass could never have pursued the extensive voyages which are now common.

Nature of Magnetism.—It is of two kinds, like the electric fluid, and every magnetic bar or horseshoe has its two poles at the extremities.

Similar poles repel; those in opposite states attract

each other. Its attraction is exercised in curves, as may be seen by scattering iron filings on a sheet of paper *placed over* a good magnet; they will form themselves into curves, of which one of the poles will be a centre.

A bar of soft iron, kept for some time in an upright position, becomes magnetic, and it may also be rendered magnetic by an electric shock.

A magnet *loses* nothing of its power by being used to rub or to attract others, but rather gains, and the power may be immensely increased by welding together a number of smaller magnets, just as the intensity of galvanic electricity is increased more by multiplication of cells in a battery, than by the size of the plates which are exposed to chemical action.

Dip and Variation of Compass.—If a small bar of steel, which exactly balances on a pivot, be magnetised, its balance is destroyed, and it will make an angle with the horizon; this is called the *dip*, or depression of the needle, and is continually changing.

As the magnetic axis of the earth does not coincide, or run in exactly the same line with the terrestrial axis, the needle does not point due north and south, but at present 24° west of north. As this direction is continually changing, it is called the *variation* of the compass.

Mariner's Compass.—The compass is generally a circular, shallow metal box, at the bottom of which is a card, in the centre of which is a point on which the needle is fixed. The card is divided into thirty-two equal parts, by lines called *rhumbs*, or points. The needle is formed into a rhomboid at each end, and over all is a plate of glass, to preserve it from the weather and wind.

Electro-Magnetism.—It was demonstrated by Oersted in 1820 that any conductor would exhibit the properties of the magnet, if it were made to form part of the circle between the poles of the galvanic battery.

If the conducting wire of a battery be coiled round a bar of soft iron, while the current passes the bar will be magnetised. This is an electro-magnet; by increasing the number and direction of the coils we increase the power of the magnet, and especially if the wire be carefully covered with silk. Bars of iron are rendered soft, as it is called, by being heated to redness, and allowed to cool gradually. Some magnets have been made so powerful, that they would support a ton weight.

QUESTIONS ON CHAPTER XXXII.

What is magnetism? When first known?
What is the *directive* power of magnetism?
How will it always point?
What is the advantage of this tendency?
In what respect is it like electricity?
How may a bar of soft iron become magnetic?
How may the power of magnets be increased?
What is the dip of the needle?
How does the magnetic needle exactly point?
Why does it not point due north?
What is the variation of the compass?
Describe the mariner's compass.
What is an electro-magnet?
How may the power of it be increased?
What is a bar of soft iron?
What is said of the power of magnets?

GEOLOGY.

CHAPTER XXXIII.

GEOLOGY (from *ge*, the earth, and *logos*, a discourse) is the science which treats of the internal *structure* of the earth, and the *materials* of which it is composed.

Those who are content with what they see on the surface, may think that the various substances which belong to the crust of the earth are arranged without any regard to order or design, but the fact is quite otherwise. Where no artificial or natural disturbance has been made, these materials are always found in a certain order of succession.

Earths.—The outer crust is composed of various earths, which consist either of decayed vegetables and plants, or of crumbled or *disintegrated* rock of some kind. Hard rocks, when exposed to the weather, are gradually disintegrated, or crumbled into small pieces. The chief are *Silex*, or flint, which produces *siliceous* earth, or sand; *Calx*, or limestone, which produces calcareous or chalky earth, and *Argilla*, or clay, which produces the clayey earth, of which pottery and bricks are made.

No one of these alone is sufficient for the purposes of cultivation. Clay alone would not allow the water to drain through it, because its particles adhere so closely together; chalk or silex alone would fail, for the opposite reason; the water would filter through them so quickly, that they would nearly always be dry. Each of the earths in its separate form is of great use in other respects.

Earth Uses.—Silica is the basis of the glass manufacture, and calx in the manufacture of soaps, sugar, &c., as well as being essential to the builder, in the various forms of lime and cements.

In some cases, as in the building of houses, all these materials are required, and at times the farmer uses chalk and sand to warm or to lighten the stiff clay soils of low-lying districts.

Proportions.—It is found that these earths are to be found in the crust of the earth in the following proportions: silex, or flint, about one half; argilla, or clay, one sixth; and calx, or lime, one eighth.

Other simple minerals, such as mica, hornblende, spar, talc, chlorite, and iron in its various forms.

All the quartz rock, and all the sand on the seashore is silex; all the shells and corals, and nearly all the bones of animals are calx.

Lime is seldom found pure, because it has so great an affinity or liking for carbonic acid, that it is almost always in union with it, and is called carbonate of lime.

Rocks.—When we have dug through the earths, we come to harder materials, called rocks, and lying usually in beds or *strata.* These are of materials similar to those which compose the earth, but in a compact or solid form.

From the accompanying diagram will be seen the various positions of the rock strata or beds, in the order in which they are invariably found, unless disturbed by certain convulsions, to be noticed hereafter. Natural examples of this may often be seen in sea-side cliffs, in railway cuttings, and steep embankments.

GEOLOGY. 117

SUPERFICIAL OR POST-TERTIARY FORMATION.		Vegetable Soil. Clay, Sand. and Gravel. Diluvial Clay, with Boulders.
TERTIARY STRATA.		Sandstone and Grits. Marls, Imperfect Limestone. Gypsum. Blue and Plastic Clays, Marls, and Lignites.
SECONDARY FORMATIONS.	Chalk.	Chalk Beds with Flints or without. Green Sands and Gault.
	Saliferous Marls. Oolites.	Wealden Clay, Limestones, and Sand. Oolitic Limestone & Grits. Lias Limestone & Shales. Saliferous Marls. New Red Sandstone. Magnesian Limestone.
	Coal Producers.	Coal Beds, alternating with Sandstone, Shale, Ironstone, and impure Limestone. Carboniferous Limestone. Quartzose Sandstone. Old Red Sandstone.
TRANSITION ROCKS.		Silurian Limestones. Grauwacké Rocks and Sandy Slate.
PRIMARY OR, METAMORPHIC ROCKS.		Clay, and other Slates. Mica, Talc, & other Schists. Gneiss, Quartz, and Crystalline Limestone.
Granite and other Plutonic Rocks.		

Unstratified Rocks.—Under the lowest strata of rocks we come to those of igneous formation, sometimes called *Plutonic* rocks, and sometimes *volcanic*, because the materials of which they are composed have been evidently fused by heat, or mixed together while in a fluid state as lava is when poured from a volcano. On this account Humboldt and other geologists called them rocks of eruption.

Unlike those which lie above them, they have no strata, while all that lie above the Plutonic have evidently been slowly deposited from above, and in a fluid state also.

The Plutonic rocks consist of granite, trap rock, and volcanic rock. Of these granite is the lowest and hardest, and forms a very durable but very expensive building stone. Trap rock appears to be more recent than granite, and the volcanic rock more so than the other two.

The trap rock, so called from *trappa*, the Swedish word for a stair, is found in great abundance in Northern Europe, often in a series of stairs or projections, and sometimes in the form of columns or pillars, as at the Giant's Causeway, in the north of Ireland, and Fingal's Cave, in Staffa.

The various kinds are called basalt, greenstone, claystone, porphyry, and amygdaloid (almond-like), which contains fragments of various stone and mineral, scattered about it like plums in a pudding, or almonds in a cake.

Volcanic rocks are less crystalline and compact, and are called lava, obsidian, scoriæ, tufa, and pumice stone.

QUESTIONS ON CHAPTER XXXIV.

What is geology?
Of what is the crust of the earth composed?
What happens when rocks are exposed?

What are the chief earths?
In what proportions are they found?
Of which are pottery and bricks made?
Of what is silica the basis?
In what manufactures is calx used?
What use does the farmer make of chalk?
How is lime usually found?
How do the rocks lie?
Where do we see proofs of this?
What is found under the stratified rocks?
Describe the nature of Plutonic rocks.
What does Humboldt call them?
In what state have strata been deposited?
Which is the hardest known rock?
What is trap rock, and where is it found?
What is amygdaloid?
To what kind do lava and pumice stone belong?

CHAPTER XXXV.

Primary Rocks.—The primary rocks, which are the lowest stratified rocks, are so called because they contain no animal or vegetable remains, and are therefore supposed to have been deposited before the existence of plants or animals.

They do not contain any fragment of other rocks, are very hard and compact, and are slaty and crystalline in appearance, which is thought to be the result of the intense heat to which they were at some time subject, through contact with the Plutonic rocks.

The principal of these series are gneiss, which is much like granite, mica, talc, and other schists or plates, quartz rock, crystalline, limestone, and clay slates.

Metamorphic.—In consequence of the crystalline appearance and texture of these rocks, some geologists have called them metamorphic, and they have doubtless been to some extent changed by intense heat.

Transition Rocks.—The series above the primary contains a few fossil remains of plants, animals, or zoophytes, and marine shells, and as they are therefore thought to have been deposited between the primary and the secondary, they are called transition rocks.

These consist of thick beds of sandstone, shale, or hardened mud, slate, and limestone, and are known as Cambrian, Silurian, or Greywacké strata.

These abound in Shropshire, the country of the ancient Silures, in Wales, and in Scotland, hence their names. They contain quantities of silver, copper, and lead, in Wales and Scotland; blacklead, in Cumberland; and quicksilver in Spain.

Secondary Rocks.—The rocks of this formation include four great systems, and numerous sub-divisions or groups. The great systems are the CARBONIFEROUS, or coal-producing; the SALIFEROUS, or salt-bearing; OOLITIC, or egg-like; and the CRETACEOUS, or chalky systems.

The Carboniferous group includes the old red sandstone, the mountain limestone, and the various seams or beds of coal, alternating with shale, ironstone, sandstone, and impure limestone.

The Saliferous system includes magnesian limestone, the new red sandstone, and the shell limestone and marls.

The Oolitic system includes the lias, or stratified limestone, the oolitic limestone, and the wealden clay; and under the Cretaceous, or chalk system, the lower green sand, gault, upper green sand, and chalk.

Unstratified Rocks.—Seeing that granite, mountain limestone, and others, lie so deeply buried under the surface of the earth, inquirers will ask how it is that in Scotland and many other places, these rocks form high mountains, towering far above the strata which are said to lie above them.

This state of things is caused by the bursting forth of the igneous rocks, through the overlying stratified beds, a convulsion which must have occurred, or we should *never see granite* and other of the unstratified rocks in Scotland and elsewhere, where whole mountain ranges are composed of that material. But though the granite has been driven through the upper strata, if the granite were taken suddenly away, the various beds would come into line again, and the one side of the strata would match the other side.

Unvarying order prevails through all the series as to position. We never find chalk below coal, nor coal below slate, and so on.

There may be a series in which some rocks are absent, as is the case where coal or chalk are sometimes found lying immediately upon slate, without the intervention of sandstone or limestone. But never, in any case, is the order found reversed, so that a geologist would never seek coal below old red sandstone or Greywacké.

QUESTIONS ON CHAPTER XXXV.

Where are the primary rocks found?
Give some description of them.
By what is their crystalline appearance caused?
Name the principal primary rocks.
By what other name are they called?
What series lies above the primary?
What remains does it contain?
Of what do transition rocks consist?
What are the Silurian, and where found?
What minerals do they contain?
What systems form the secondary rocks?
What are the carboniferous rocks?
To which do chalk and marls belong?
Explain how granite is found at the surface.
What is said of the order in which rocks lie?
Where is granite found in abundance?

CHAPTER XXXVI.

Tertiary Formation.—Above the chalk system we find three groups of rocks in regular succession, called Eocene, Miocene, and Pliocene, and which are chiefly deposited in basins or hollows, being evidently of more recent formation than those of which we have spoken previously.

Eocene.—The Eocene, or lowest of the Tertiary strata (from *eōs*, dawn, and *kainos*, new or recent), is so called because the remains of animals which are found in it are similar to those of animals which now exist. It is therefore called the dawn of a new period. Out of one hundred species of fossils, nearly four per cent., or four in every hundred, are like those of existing species.

Miocene.—In the Miocene (from *meion*, less) this number increases from four per cent. to seventeen.

In the Pliocene, or more recent (from *pleios*, more), the number of extant species amounts to nearly fifty per cent., and in the upper strata, or new pliocene, more than ninety per cent., that is, out of one hundred fossil animals, ninety or more are similar to those which now exist.

Proofs of Design.—There are persons who teach that all creation is governed by natural laws, which operate of themselves. Geology, which is sometimes called the "Great Stone Book," teaches that there was a time when man did not exist, and that other animals lived before him. Man must therefore have been created, as the Book of Genesis declares, after other animals. His Creator must be self-existent, uncreated, everywhere present, acting and ruling. Such a creator is God alone.

GEOLOGY. 123

The Tertiary rocks consist of *marls*, a composition of clay and chalk, in which there is more clay than chalk, gypsum, or plaster of Paris, loose limestone and sandstone, lignite, or half-formed coal, blue and other plastic clays.

Eocene beds were found in England and France, *miocene* beds in France, Germany, and other parts of Europe, but not in England, and *pliocene* in various parts of Great Britain and Ireland. London and Paris are built on eocene basins, and consequently there is an abundance of good clay in the vicinity of the one, while gypsum is found in great quantities in the neighbourhood of Paris.

Eocene Fossils.—In this system it is very remarkable that more than 1200 species of fossil shells have been found, only about 40 of which are now known to exist.

Of forty large quadrupeds, the remains of which have been found, and some of which are as large as the horse and the rhinoceros, there are only four living species at present in existence, and of these three are of the tapir tribe. Some of large size have been found in the Isle of Wight. From the great size of these remains, it is evident that the vegetation of the part of the world which they inhabited was of a tropical character, as both monkeys and palm trees have been found, as well as fossil elephants, hippopotami, and others of the pachydermatous, or thick-skinned animals, and turtles.

It is also seen that the beds of lignite contain remains of such vegetation as would be required by creatures of that growth.

Miocene Fossils.—There are none of these in England, but at Epplesheim, in Hesse Darmstadt, Germany, numerous very large fossils have been discovered of an animal similar to the tapir, and the largest of all which has yet been found. It has been called the dinotherium. The head alone measures four feet in length, and three in breadth, and is furnished with huge tusks bent downwards.

Pliocene Fossils.—The pliocene fossils are found in the glacial drift of Norfolk, in North Wales, and on the Clyde, in Scotland. There are also specimens found in Yorkshire, notably in a cave at Kirkdale, where the bones of 300 hyenas were found, as well as those of the elephant, hippopotamus, horse, bear, wolf, and several birds, in all more than thirty different species.

QUESTIONS ON CHAPTER XXXVI.

What rocks are found above the chalk?
Of how many groups do tertiary rocks consist?
Which is the lowest? What does it contain?
Why is the eocene system so called?
What is peculiar to the pliocene group?
Of what do the tertiary rocks consist?
Where are eocene beds found?
Which are not found in England?
What is said of London and Paris?
What is gypsum? Where is it found?
In what groups have large quadrupeds been found?
In what island are some of them found?
What is said of the vegetation of the time?
Where are miocene fossils found?
What are the largest of them called?
Where are the pliocene fossils found?
What were found in Kirkdale Cave?

GEOLOGY. 125

CHAPTER XXXVII.

THE following tabular arrangement, which is generally agreed upon by geologists, will be useful for reference.

There are ten systems, as follows:

I. The **Post-tertiary** system, consisting of all earthy deposits, peat-mosses, coral-reefs, raised beaches, and containing remains of plants and animals of *existing species*, or of species recently extinct, or no longer existing.

II. The **Tertiary** system, embracing all the regularly stratified clays, marls, limestones, and lignites, which are found *above the chalk*, and containing remains of plants and animals mostly extinct, but *similar to existing species*.

III. The **Cretaceous, or Chalk** system, comprising all the chalk and greensand beds, and containing remains of plants and animals, *chiefly marine, and all extinct*.

IV. The **Oolitic**, consisting of the Wealden, oolitic, and lias strata, and containing plants and animals *chiefly reptiles, and all extinct*.

V. The **Triassic, or Triple** system, including (1) the saliferous marls, (2) muschelkalk, or shelly limestone, and (3) the variegated sandstones, with remains similar to the oolitic.

VI. The **Permian**, including lower portion of the new red sandstones and magnesian limestone, with plant and animal remains, *similar to those found in the carboniferous system*.

VII. The **Carboniferous** system, embracing the coal formations, mountain limestone, ironstones, and carboniferous slates. It has abundant remains

11—3

of plants and animals, the plants being nearly all *endogenous* (see Botany) in the coal, and the animals in the limestone *all marine shells, fishes, and zoophytes.*

VIII. The **Old Red Sandstone, or Devonian** system, including the yellow sandstone, red conglomerate, and grey flagstone groups. Some remains of fishes, crustacea, and shell-fish, *but very few plants.*

IX. The **Silurian** system. It contains numerous remains of *marine invertebrate animals.*

X. The **Primary, or Metamorphic** system, containing all hard and crystalline rocks, *and quite destitute of fossils.*

These have also been arranged in four periods, in accordance with the fossils found therein :

I. The **Cainozoic,** or those of recent life.
II. The **Mesozoic,** or those of middle life.
III. The **Palæozoic,** or those of ancient life.
IV. The **Azoic,** or those void of life, or destitute of fossils.

QUESTIONS ON CHAPTER XXXVII.

How many systems does the tabular arrangement contain?
Name the first and the tenth.
What remains does the post-tertiary contain?
What rocks are included in the permian?
What system is found above the chalk?
Which contains extinct animals, but similar to existing species?
Which contains marine animals all extinct?
What rocks are included in the triassic system?
In which system are coal and iron found?
What plants are found in the carboniferous system?
What system is quite destitute of fossils?
In what formation are huge reptiles found?
What is peculiar to the metamorphic rocks?
Name the four fossil periods.
What is meant by the *azoic?*

HEAT.

CHAPTER XXXVIII.

Definition.—Heat has been defined as "the essence that gives warmth, and which preserves gases and liquids from becoming solid."

Heat which is felt on approaching a fire is called *sensible*, that which has to be produced by mechanical or other means is called *latent*. All bodies possess latent heat, however cold they may seem to be.

Sources of Heat.—The chief source of heat is thought to be the earth itself, which many geologists suppose to be a globe of burning matter. If we descend far below the surface, as in mining, or even well-sinking, we experience an increase in the temperature: it is warmer.

If, as some believe, the temperature increases at the rate of one degree for every fifty feet of descent, it is certain that the interior must be in a fused or melted state, the cool parts of which are the comparatively thin crust of the earth, and which is occasionally broken through by the combustible matter, or accumulated gases below, as in the case of earthquakes or volcanoes.

Caloric.—Another source of heat is the sun, but heat, or *caloric*, as it is called in science, may be developed from almost everything. As we have seen in chemistry, heat is produced by chemical action in the mixing of some cold liquids, as water and sulphuric acid, or a liquid and solid, as water and quicklime.

Friction.—Caloric is also produced by the friction of solids. The Indian will procure fire by rubbing pieces of wood together; the school-boy heats a button by rubbing it on wood; so great a heat is caused in boring cannon, that a continuous supply of oil or water is required to prevent the metal from softening. For a similar reason oil or some other fatty substance is required to be applied to the wheels of railway or other carriages, to prevent them from taking fire through the excessive heat which rapid motion would cause. The friction of fluids, such as oil and water, does not produce heat.

Conduction of Heat.—Some bodies *conduct* heat better than others; iron and other metals better than wood or glass; a piece of burning wood, or red-hot glass may be held much nearer the heated part than a similar heated rod of metal could be. For this reason the handles of tea and coffee pots should be made not of metal, but of wood, ivory, or some other bad conductor.

If a sheet of paper be wrapped tightly round a metal rod, and held in the flame of a spirit lamp, the paper will not take fire, because the metal carries the heat away so rapidly. If a wooden rod be used instead of metal, the paper will soon take fire.

Dress.—By wrapping ourselves in various kinds of clothing, we keep the caloric from passing out of our bodies, and thereby we are kept warm. By a series of experiments it has been shewn that a thermometer which was cooled in the air in 27 seconds, required more than 1000 seconds to cool when wrapped in cotton wool, while other material required a longer time. increasing as follows: sheep's wool, raw silk,

HEAT. 129

beaver's fur, eiderdown, and hare's fur, the last of which caused the time of cooling to be 1315 seconds.

Roofs.—Thatched cottages are warmer in winter, and cooler in summer than those covered with tile or slate. In winter the warmth is not conducted *outwards* so quickly, and in summer the heat is not so quickly conducted within, as straw is not so good a conductor as slate and tile. For a similar reason, snow on the ground keeps the soil warm.

Heat Ascends.—We place water above a fire to heat it; the lower portion becomes warm and ascends, while the cooler descends and takes its place, until all becomes similarly heated, and when the heat has accumulated to 212° Fahrenheit, the liquid boils. There are continuous *currents* all the time the water is being heated.

The same phenomenon takes place when air is heated. That which is over the equator is greatly rarefied, and rises, cooler air rushes in to supply its place, and thus winds more or less powerful are caused.

Ventilation.—The ascent of the heated air enables us to ventilate buildings, ships, mines, &c. A mine is ventilated by keeping up a great fire at the bottom of it; the heated air ascends, and other cold air rushes down the mine to supply its place.

Buildings have openings of various kinds in the roof to let out the heated air, but the whole subject requires to be better understood by those who build houses and ships.

In steamships the lower part of the vessel is kept ventilated by the draught of the boiler furnaces.

QUESTIONS ON CHAPTER XXXVIII.

How is heat defined?
Distinguish between sensible and latent heat.

130 OUTLINES OF SCIENCE.

What is said about mine-sinking?
What are the chief sources of heat?
What is the scientific name for heat?
How is heat produced by chemical means?
What examples are given of friction?
Which are the best conductors of heat?
How is wood shown to be a bad conductor?
What is the chief advantage of dress?
What is said of thatched cottages and of snow?
What happens when water or air is heated?
How is water heated? What is boiling heat?
What enables us to ventilate ships?
How is a mine ventilated?
Why is a steam-vessel better ventilated?

CHAPTER XXXIX.

HEAT—continued.

Radiation.—Caloric passes from a heated body in straight lines, and in all directions, as does light—this is called radiation. Dark and rough surfaces radiate heat, and also absorb it most quickly and perfectly. Whatever is required to retain heat, such as teapots, boilers, &c., should be as light coloured and as smooth as possible.

Absorption of heat takes place generally when it passes through any substance. The darker the colour of the substance, the greater is the amount of heat absorbed.

Reflection.—Heat is readily reflected by plates of polished metal, but there is a great difference in their reflecting power. A plate or reflector of polished gold will reflect much more than a plate of brass, and more than twenty times the amount of heat that a blackened metal plate will reflect. The angles of incidence and reflection are equal, as in the case of the reflection of light.

HEAT. 131

Expansion and Contraction.—All bodies, whether solid, gaseous, or fluid, are expanded by heat. There are two or three apparent exceptions, to be hereafter noticed.

If a piece of iron which exactly fits a ring when cold, be made hot, it will not go into the ring until it becomes cold again. Clocks and watches, which are not specially adapted for a variable climate, are liable to gain or lose according as the weather becomes cold or hot; hence in warm weather we find it necessary to shorten the pendulum, and in cold weather to lengthen it in the same proportion. Cast-iron pipes are lengthened about six inches in 300 yards by the change from winter to summer.

Now that so much iron work is used in buildings, it is of the utmost importance that this should be allowed for, and it is so in the Menai, and other large bridges, as well as in other structures which are superintended by really scientific engineers.

Iron tires are applied to wheels when very hot, so that as they become cold they contract, and grip the parts of the wheel more closely.

Contraction Applied.—It happens occasionally that the walls of a building get out of the perpendicular, so as to become unsafe. This was the case with a large museum gallery in Paris, and to restore the inclined wall to its upright position, bars of iron were passed from side to side, and screwed tightly by nuts on the outside. After screwing them to the utmost while cold, the bars were heated, and while expanded, the nuts were screwed up more tightly than before. The result was, that as the heated bars became cold, they contracted again, and by a silent but irresistible force drew the wall upright.

Sudden Expansion.—If a quantity of sulphuric acid be carelessly poured into a glass vessel of cold water, the heat generated will break the vessel, especially if the glass be thick. Hot water will often have the same effect, and glass may be broken by running a red-hot wire over a plate of it, when it may be as easily cracked off as though it had been starred with a diamond.

The reason is that the inner surface expands much more rapidly than the outer, and is torn from the outer by the sudden expansion.

Ice and Clay.—Water does not continue to contract by cold below the temperature of $40°$ Fahrenheit. Below $40°$ as the cold increases, it expands, so that when it becomes ice, its specific gravity is less than that of water, and it rises to the surface and floats.

Design.—By this wise and beneficent arrangement, rivers are prevented being permanently frozen; were it otherwise, the whole of the water would become a mass of ice, and the living creatures in it would perish.

The wonderful expansive force of small quantities of water may be seen by the breaking off of rocks which have been bored, and water poured into the hole during a frost. The moisture absorbed by earthy matter in the autumn, frequently causes landslips during the ensuing frosts.

An unpleasant illustration of expansive force is seen in the bursting of our water bottles and jugs by a severe frost. At such times they should be emptied when the frost begins.

Clay, which is commonly spoken of as an exception to the rule as to expansion by heat, takes up less space after it has been baked, but the reason doubtless is that it parts with much of its water in the process of baking, and therefore becomes substantially less. Clay, like other earths, is broken up, and its particles loosened, by the effect of frost.

The expansiveness of metals is generally in accordance with their fusibility. Those which are easily fusible, such as lead, are also very expansive.

HEAT. 133

QUESTIONS ON CHAPTER XXXIX.

What is radiation?
What colours and surfaces radiate most quickly?
What occurs when heat passes through a substance?
What metal has the greatest reflecting power?
How are all bodies affected by heat or cold?
Give examples of expansion by heat.
How do expansion and contraction affect buildings?
In what condition are the tires put on wheels?
What remarkable case of contraction is mentioned?
What occurs when sulphuric acid and water are mixed?
Give some reason for the breaking.
What occurs to water below 40° Fahrenheit?
What would happen if it kept on contracting?
How are rocks sometimes split?
Name another exception to the rule.
What is said of the fusibility of metals?

CHAPTER XL.

HEAT—continued.

Thermometer (from *thermos*, heat, and *metron*, a measure) is an instrument for measuring the degrees of heat in the atmosphere by expansion, the invention of which has been ascribed to various persons.

The ordinary thermometer consists of a glass bulb, having a fine stem; the bulb and stem are filled to a certain height with alcohol or mercury, and in the rest of the tube there is a vacuum.

By subjecting the mercury in the tube to a great heat, it expands, and fills up the small tube, when it is carefully sealed up, and allowed to cool. It is then graduated, so that we may reckon the degrees of heat or cold. When plunged into boiling water for a short time, the water rises to boiling point, 212° Fahrenheit. If placed in snow or melted ice, it falls to 32°.

The Centigrade and Reaumur Thermometers.—The thermometer used in France is called Centigrade. Its freezing point is marked 0°, while its boiling point is 100°.

That used in Germany is called Reaumur, which also has its freezing point 0°, and its boiling point is only 80°. The centigrade is the one most in use on the continent of Europe.

Evaporation.—Most fluids, and many solids, constantly give off vapour. When this is done without disturbing the surface, it is called Evaporation. When the surface is disturbed it is called **Ebullition** or boiling. Heat is the cause of both.

Evaporation produces cold, hence in India, and other hot climates, water is poured or sprinkled freely on floors, furniture, roofs, tents, &c., because its evaporation lessens the heat.

Various liquids, such as prussic acid, will freeze by evaporation. Thus if ether be poured upon water, and placed under the receiver of an air-pump, when the air is exhausted the ether will boil away, the heat will be evaporated, and the water will freeze.

The Hygrometer.—This is an instrument, various in form, made to measure the quantity of water in the atmosphere. As all vegetable fabric absorbs moisture, it thickens and shortens in the process, and is sometimes used as a hygrometer.

The ropes in a drying-ground may be tightened by wetting them, and on one occasion, while raising the famous Egyptian Obelisk in front of St. Peter's, at Rome, this fact was usefully illustrated. When it was found that the ropes were a little too long, so that the work could not be completed, some one in the crowd shouted out, " Wet the ropes." This was done, and the result was that the ropes contracted so much, as to raise the obelisk to its vertical position.

Dew.—Part of the moisture which is evaporated from the surface of the earth during the day, is deposited there again during the night in the form of dew. Vegetable fibre, and similar substances which radiate heat rapidly, are the first recipients of the dew, consequently the leaves of trees, grasses, &c, are frequently refreshed by it.

Radiation and dew-deposit are both more rapid when the sky is clear; in cloudy weather the vapour is driven back to the earth.

Rain.—When the moisture which has been evaporated rises into colder regions, it becomes dense and forms clouds.

If the air be disturbed by electricity or other causes, the vapour forms into round drops and becomes too heavy to float. It then falls in the form of rain.

Snow and Hail.—Sometimes, as these drops fall, they pass through a very cold current of air, and are frozen into little lumps of ice as hail, or into less solid masses as snow.

Freezing Mixtures.—If salt and snow be mixed together, the snow will be melted, and a thermometer placed in the fluid will fall to zero. Solid carbonic acid and sulphuric ether, when mixed, form the coldest known substance.

A mixture of crystallized chloride of calcium with snow produces the coldest known temperature. The mercury of a thermometer placed in it, will sink to 66° below zero.

Other freezing mixtures are snow and diluted nitric acid, or snow and potash, which are easily obtained. The solid carbonic acid can only be obtained from carbonic acid by a pressure of thirty atmospheres, 450 lbs. to the square inch, which reduces it to a liquid, and that liquid becomes a solid at 94° below zero.

OUTLINES OF SCIENCE.

QUESTIONS ON CHAPTER XL.

How can we measure degrees of heat?
Describe the thermometer.
What thermometer is used in France?
What is the difference between the Centigrade and Reaumur?
What is evaporation? What ebullition?
How is heat lessened in hot countries?
What effect has evaporation on ether?
What is the use of the hygrometer?
How is vegetable fibre affected by moisture?
Give some examples of rope-shrinking.
What is dew? When is it most heavy?
What articles are most benefited by it?
How is dew affected by cloudy weather?
Explain how rain is formed.
What is hail, and how formed?
What articles form the coldest substance?
What mixture sends the mercury down to 66° below zero?
How is solid carbonic acid obtained?
What is its temperature?

PHYSIOLOGY OF MAN.

CHAPTER XLI.

The bodily constitution of man is one of the most important to all, but one of the least understood by the majority of persons. The health of the mind depends largely upon that of the body, while the latter is under our control to a much greater extent than is generally understood.

Man can never be cheerful and happy, as a rule, except when in health. This applies especially to young persons, whose mental faculties are quickly influenced by the disorder of the body.

Suitable food, exercise, and cleanliness, pure water, and good air, are essential to perfect health and vigour.

Waste of the Body.—It has been proved by experiment that man cannot take exercise, or use any bodily force, without some waste of the material of which the body is composed. It is quite possible indeed to lose some of its particles in a state of absolute quiet or idleness, for what is called the insensible perspiration would still go on.

The chief process by which this waste is repaired, is by taking wholesome food; it is not the only means, because, as we have said above, pure air is required to assist in this process.

Waste Repaired.—Man must therefore eat to live, and it is well that young people should understand how the digestion and assimilation of the food is accomplished, what delicate machinery is in operation to keep them in life and health, and how wisely and beautifully all has been designed by the Creator for the happiness of his creatures.

A tolerably accurate knowledge of the bodily formation and functions will go far to prevent the blunders and excesses to which young people are especially liable.

The Bodily Frame.—The great internal supporter of the body is the bony skeleton, which contains more than two hundred bones. The largest of these are the vertebræ of the backbone, which in early life contains thirty-three separate bones, some of the lower of which, however, unite as we grow old.

The Spinal Marrow.—The backbone contains the spinal marrow in a hollow groove in its centre, and at the top of the vertebræ the skull, formed of numerous separate bones, some of which unite as we come to manhood, and which contains the brain.

From the marrow and brain the nervous system extends all over the body like the numerous branches of a spreading tree, and wherever pain can be caused there is a nerve.

The Ribs.—Connected with the vertebral column or backbone are the ribs, twelve on each side, eighteen of which are usually attached to the sternum or breastbone, and within which the most important soft parts of the human frame are contained and protected from injury, such as the heart, lungs, and other vital organs.

The Limbs and Synovia.—The legs are joined to the pelvis, which comes between them and the backbone. Each leg is composed of thirty bones, and each arm has the same number, and these are fastened together very strongly by a tough substance called ligament or cartilage, and where one bone plays or moves inside another, as in the joints, the ends which come in contact with each other are not only lined with a soft smooth cartilage, but within each socket an oily fluid is secreted or formed, called the *synovia*.

This acts on the joints like oil on the wheels of the steam engine. When by accident or other means this joint-oil has been interfered with or allowed to escape, the result is always a stiffness of the limb.

The Muscles.—A mere skeleton would not be able to maintain the upright position, although the feet of man are admirably formed for that purpose. A series of strong bands of flesh called muscles are necessary, whith under the influence of the will contract or shorten, and pull the various parts of the body in different directions.

The chief of these are the muscles of the spine, which keep the body from falling forward, the muscles of the thigh, and those of the legs. These are lapped over each other so regularly, that an anatomist can separate one from the other, and each has its proper name and office.

The Skin.—Over the muscles we find the skin, which is much more complex than it appears to be from the outside. It consists of two parts, the *dermis* and the *epidermis*—the dermis lies immediately over the muscles, and is very sensitive of pain or injury; the epidermis, or outer skin, does not feel pain, so that you may remove a portion of it without suffering.

The epidermis, or cuticle, does not absorb anything easily, otherwise we should be always in danger of blood poison; but now we can handle poisons and even putrid matter without danger, so long as the skin remains whole.

Blood Poison.—It is quite otherwise with the dermis, which is so easily poisoned or inflamed, that anatomists, in operating on dead bodies, have sometimes died from an accidental cut or scratch made by their own dissecting knives.

Vaccination.—So in vaccination or inoculation the virus is introduced into the blood-vessels of the dermis, after cutting through the cuticle.

The skin performs an important function or duty, as it is proved that an acid called urea, as well as carbonic acid and watery vapour, are filtered outwards through the million of pores which are spread over it like the meshes of a net, and through which it is computed that about eighteen ounces of water, and nearly an ounce each of carbonic acid and solid matter passes every day.

Men who are employed at gas works, or on board large steamers, afford examples of the loss of material through the skin. Such men, who have been weighed on going to work, have been found to lose from four to five pounds weight in the course of as many hours.

Between the cuticle and the true skin there is a layer of colouring matter, which in the negro race is black, and varies according to the complexion of the various races.

QUESTIONS ON CHAPTER XLI.

Why is the study of physiology important?
What is the result of every bodily movement?
How is the waste repaired?
What is the great supporter of the body?
How many form the backbone?
What does the backbone contain?
How many in each arm and leg?
To what is the nervous system compared?
In what are the heart and lungs enclosed?
How many ribs are there in each side?
What is the synovia? What is ligament?

What is the function or duty of the muscles?
What are the dermis and epidermis?
Show how the epidermis is fitted for its duty?
What is the special duty of the skin?
What is found between the two skins?

CHAPTER XLII.

THE DIGESTIVE PROCESS.

Food, and the Mouth.—Suitable food being provided, it must be well masticated or chewed, during which it is mixed with a fluid called the saliva, or spittle, which is secreted in the mouth, by six little glands. This saliva acts chemically on some kinds of food; it will change the starchy elements of the food, which are not nutritious, into sugar, which is highly so. It moreover prepares the food for easier digestion in the stomach, and therefore thorough mastication is important.

The Teeth.—The teeth, which perform so useful a purpose, are thirty-two in number, of three different forms, and adapted to all kinds of food. They are faced with a hard enamel, and as the back of the tooth is softer than this enamel, the tooth wears down with use, and always maintains a sharp edge, and this especially in the incisors, or cutting teeth.

When the food has been thus masticated and insalivated, it is ready to be swallowed, and passes over the tongue into the gullet or esophagus, down which it goes into the stomach.

The Stomach and its Duties.—When the food passes down the gullet into the stomach, that organ contracts and moves it about, thus mixing it with a fluid which is secreted by numerous glands in its coats, and which is called the *gastric* juice.

The acid or dissolving power of this fluid is very great, and it speedily reduces the solid food to a pulp, after which it is allowed to pass through the pylorus into the duodenum or large bowel, in the state of chyme. Part of the liquid mass is believed to be absorbed through the sides of the stomach, and carried direct into the bloodvessels which are scattered over it. The pylorus is the lower opening of the stomach.

Time of Digestion.—By a singular circumstance, Dr. Beaumont of America was enabled to make experiments of the time in which various kinds of food would be digested. A young Canadian, named Alexis St. Martin, had received a gunshot wound in the stomach; the wound healed round the edges, without closing, and left a hole, through which the contents of the stomach could be seen.

Shortly afterwards a natural valve was formed over the inside of this hole, which could be pushed inwards, but which would not allow the contents of the stomach to escape.

Experiments.—Dr. Beaumont hired him for the sake of performing experiments, and the result of frequent trials proved that various kinds of food were digested in about the following spaces of time:

Fibrinous food, such as roast beef and mutton, in 3 hours; salt pork, 4½ hours; strong beef soup, 5 hours.

Albuminous food.—Eggs, raw, 1 hour; soft boiled, 3 hours; hard boiled, 5 hours; boiled fish, 2 hours; fried fish, 3 hours; raw oysters, 2 to 3 hours.

Gelatinous food.—Tripe, 1 hour; boiled veal, 4 hours; roast chicken and sucking pig, 2 to 3 hours.

Fatty food.—Broiled fat pork, 3 to 4 hours; fried bacon, 5 hours.

Cheese-like food.—Boiled milk, 2 hours; raw milk,

nearly 3 hours; toasted cheese, 3 hours; raw cheese, 4 hours.

Farinaceous food.—Boiled rice, 1 hour; boiled potatoes or beans, 2 hours; baked wheaten bread, 3 to 4 hours.

Mucilaginous food.—Boiled carrots, parsnips, or turnips, 3 hours.

It must be noticed that when the stomach is overladen with a heavy meal, the process of digestion is slower, because the gastric fluid cannot so easily mix with all the particles, to reduce them to chyme. An ordinary meal of mixed food is so reduced in about 4 hours.

Chyme changed to Chyle.—When the food in the state of chyme passes into the duodenum, which is the part of the intestinal canal next to the stomach, it receives two other additions, the bile from the gallbladder, and a secretion from the pancreas or sweetbread, which lies behind the stomach.

The Bile.—The alkali of the bile neutralizes the acid of the chyme, the starch in the food is converted into sugar, and the whole nature of the fluid is changed. It is now called chyle, and is in a condition to be assimilated, or used to repair the bodily waste of which we have spoken above.

Assimilation.—The chyle is forced along the small intestines, which succeed the duodenum, by what is called a peristaltic, or worm-like motion, and in its passage the nutritious portions of the chyle are imbibed or absorbed by minute spongioles, in the inside of the bowels.

The Mesentery.—These are connected with the mesenteric gland, and passing along, meet at one point, and are received in a tube about the size of a little finger, and which is called the *thoracic duct*.

Nutrition.—This thoracic duct passes upward to the top of the aorta, one of the chief blood vessels to be noticed presently, and empties itself into the large vein called the vena portæ, whence it becomes mixed with the blood, and is used for all the purposes of nutrition, renovation, growth, and repair of waste. By this time the chyle has assumed a pinkish colour, and if kept still will separate, like blood, into a solid and a watery part.

Epithelium.—The whole interior of the alimentary canal, from the lips downward, is lined with a soft coating called the mucous membrane, also the epithelium, very tender and sensitive, and which is full of glands of various kinds, for the objects just mentioned.

QUESTIONS ON CHAPTER XLII.

What is mastication? Why important?
How does the saliva affect the food?
With how many teeth are we supplied? Describe them.
Whither does the food pass after mastication?
What happens to it in the stomach?
Into what is it changed by the gastric juice?
What is the pylorus? And the duodenum?
What becomes of a portion of food in the stomach?
How have we learned some facts about digestion?
How long does fibrinous food take to digest?
What is gelatinous food? How long does that require?
Name the foods which are most easily digested.
What is added to the chyme in the duodenum?
What effect has the bile on the food?
Show how the nutritious food is assimilated.
What are the duties of the mesentery and thoracic duct?
How is the alimentary canal lined?

PHYSIOLOGY OF MAN. 145

CHAPTER XLIII.
THE BLOOD AND ITS VESSELS.

Its Composition.—The blood is said by chemists to be "an alkaline fluid, consisting of water, and of solid and gaseous matter."

In its healthy state it is always at a temperature of about 100° of Fahrenheit, marked blood heat on the thermometer.

In every 100 parts of blood there are 79 parts of water, and 21 parts of dry solids, just as in air there are about 79 parts of nitrogen, and 21 of oxygen.

The globules of blood in human beings are always circular, while in other animals they are elliptical or oblong in form.

Separation.—When blood is drawn from a body, it quickly cools, and in about fifteen minutes it separates into two different constituent parts—a clear yellowish liquid called the *serum*, and a much thicker and red part called the *clot* or *crassamentum*.

The clot contains the more solid portions, called fibrin, from which the skin, muscles, and vessels are made, and which usually sinks to the bottom as the blood cools. It also contains iron, and various phosphates.

Hæmatine. — The red colour of living blood is caused by the presence of millions of red corpuscles or globules, of the shape of a flattened pea, and of which many thousands could lie on a square inch. Small as they are, they have a little ball of white matter inside, as the red matter or hæmatine is only a colouring case or envelope, and can be washed away.

The serum contains more fluid matter, and albumen. The whole constituents of the blood (26 in number) are water, hæmatine, albumen, fibrin, fatty

13

matter, iron, and very minute portions of various other chemical substances.

Lymph.—There is another fluid found in the various cavities of the body which is not coloured like the blood, but which contains fibrin, and is daily poured into the blood. This is called lymph, its vessels are called lymphatics, and so far as is known at present, it exudes through the tissues as the chyle is being assimilated in the process of digestion.

The Circulation.—Since there is a continual waste going on in the human frame, which can only be repaired as new material is supplied by the blood, it is necessary that there should be a machinery to carry that fluid to all parts of the body. Such a machinery was discovered to exist by Dr. Harvey, who lived in the reign of James I.

By this discovery he learned that blood is sent from the heart through a system of *arteries*, and that it returns to the heart through a similarly arranged system of veins.

The Heart.—The heart is the great reservoir of the blood; it is arranged in four separate chambers or compartments with very strong muscular walls between, and containing eleven valves, which only open one way, so that blood which has once passed through them, can never return the same way.

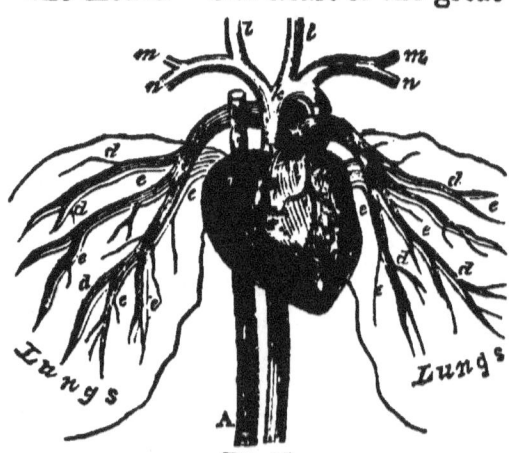

Fig. 25.

PHYSIOLOGY OF MAN. 147

From fig. 25, which is a front view of the heart, with an outline of the lungs, and the pulmonary or lung veins and arteries, it will be seen that there is an intimate connection between the heart and lungs.

Circulation.—By the muscular action of the heart, the venous blood, of a dark red colour, is forced through the right ventricle into the large pulmonary arteries, *d d*. By this it is spread through numerous branches over the surface of the lungs and air passages, and passing through exceedingly minute channels, called *capillaries*, it returns by the pulmonary veins through four valves, into a third chamber of the heart, called the left auricle. From the left auricle it passes by another valve into the left ventricle, a chamber with very strong muscular walls, from which it is pumped out of the aorta, *k*, through numerous arteries, to every part of the body.

Altered State of the Blood.—As the blood returns through the pulmonary veins to the heart, it is seen to be of a lighter colour than when it was sent out, so that it must have undergone an important change in the course of its passage through the lungs. This brighter coloured blood is sent through the arteries all over the body, and passing through capillaries at their extreme ends, comes back by the veins to the heart of a much darker colour. In its course it supplies materials for growth and renovation to all parts of the body, and evidently takes from the interior some or its impure or waste matter, which is brought back to be purified in the lungs.

Venous and Arterial Blood.—The cause of this change of colour is not thoroughly understood, but it is believed to be due to the presence of oxygen or carbon. When pure air is taken into the lungs, it comes into contact with the venous blood in the pulmonary arteries, which contains much carbonic acid. The oxygen is taken into the blood through the very thin membrane of which the lung passages are composed, and the carbonic acid passes through into the passages to be expired or breathed out of the mouth. By this process of inhaling pure air, the *venous* blood becomes *arterial*.

On this account it is that pure air is so important to health, for without it the same air has to be breathed over again, and in a short time all the pure oxygen will be exhausted.

Carbonic Acid formed.—The air is taken into the lungs dry and pure; it is exhaled again moist and abounding in carbonic acid. If any person breathe through a glass tube into a tumbler of pure lime-water, it becomes turbid or thick, thereby proving that carbonic acid gas has been forced into it. It is this carbonic acid which forms a chief ingredient in the life and nourishment of plants. They take in carbonic acid and expel oxygen. Animals use the oxygen, and expel the carbonic acid.

QUESTIONS ON CHAPTER XLIII.

How is the blood defined by chemists?
Of what is it composed? Its temperature?
Of what shape are the globules in human blood?
What happens when drawn blood cools?
What is contained by the clot?
What is hæmatine? Of what is serum composed?
What is meant by lymph and lymphatics?
How is it supposed to enter the blood?
How is the waste of the body supplied?
Who discovered the circulation of the blood?
What are the arteries? and the veins?
How is the heart divided? How many valves?
With what other organ is the heart connected?
Describe the action of the heart and lungs.
What change takes place in the lungs?
How is arterial blood known from venous?
What is supposed to change the colour?
Why is pure air of great importance?
What is remarkable in the feeding of plants?

CHAPTER XLIV.

ARTERIES, VEINS, CAPILLARIES, AND SECRETIONS.

THE *arteries* are always found empty of blood after death. They consist of three coats, the outer of which is very elastic, the middle less so, and the innermost least so of all.

The result of this is that in the case of the severance of an artery by any violent means, the inner coat of the artery is first to break, and curling up, draws the whole end of the artery into the tube, like a plug, which thus prevents the rapid escape of blood.

The arteries are nearly straight, and always well protected by bones and muscles. The veins are similar in structure, but more branched and crooked; the blood flows through them with less force, so that the wound of a vein is not so dangerous as that of an artery, nor so difficult to cure.

The Capillaries.—These are so small, that the passage of blood from the arteries to the veins cannot be seen by the naked eye.

The Pericardium.—The heart is surrounded by a bag called the pericardium, which contains a fluid, and the whole of the lungs, heart, &c., are contained in and protected by the bony cavity of the chest, and a strong membrane that lies below them, called the diaphragm.

Secretions.—The most important of these are the saliva, the gastric juice, the bile, the pancreatic fluid, lymph of various kinds, that of perspiration, and the urine. Like the action of the heart and other important organs, these secretions are carried on quite independent of the will, for example, the sight of fruit, or other palatable things, or even the thought of them, will sometimes cause the mouth to water, that is, cause the saliva to flow.

Secretion is that process by which materials are separated from the blood, and from the organs in which they are formed. They are either made use of in the body, or expelled from it as useless.

Secretion is as necessary to health as nutrition, and may be arranged in three divisions, as *exhalations*, when moisture is breathed out from the mucous membrane, *follicular* secretions, and *glandular* secretions.

Where material is discharged from the body, as in the case of perspiration, it is called an excretion. The skin is furnished with numerous small holes or pores, which are the openings of small hollow organs called follicles, with membranous sides. These organs are usually filled with fatty or albuminous matter.

Perspiration.—This is a follicular secretion of a watery vapour, which is incessantly passing off through these pores. It consists chiefly of water, but contains also acetic acid, muriates of soda and potash, and other solids.

The free flow of this secretion is essential to health, for whenever the pores of the skin are closed, disease of some kind immediately ensues. Hence cleanliness and exercise are very important, as tending to promote its free action.

Glands.—Glandular secretions are of different sorts ; the chief are the tears, saliva, bile, milk, pancreatic fluid, and the urine.

Saliva.—Of the saliva and its uses we have spoken elsewhere, as being of great importance in changing the starch of the food into sugar, and so rendering easier the process of digestion.

Tears.—These consist of the limpid fluid secreted by what are called the lachrymal glands. The object of tears is to moisten the cornea, and to prevent friction, by flowing over the surface of the eyes. They are also useful in washing off foreign substances from the eyes.

Bile.—This oily fluid is secreted by the liver; it is of a yellowish-brown colour, and has a bitter taste. The bile is separated from the blood by the liver, and when not required for digestive purposes, it is stored up in a hollow sac or bag, called the gall-bladder.

The use of the bile appears to be to change the digestive food from chyme to chyle. It is also thought to assist in causing the peristaltic or worm-like motion of the bowels, as a deficiency of bile is usually accompanied by a torpid action of the bowels.

Pancreatic Juice.—This fluid is secreted by an organ which is common to all the vertebrate animals, and which is known as the sweetbread. It is similar to the salivary glands, but softer, and the liquid which it secretes is like the saliva in appearance, and assists in the process of digestion.

Urine.—This is a fluid which is separated from the blood by the action of the kidneys, to the extent of thirty or forty ounces per day, according to the quantity of fluid drunk. Whenever the action of the skin is deranged, the kidneys are affected; when perspiration ceases they are overworked, and often become diseased in consequence.

It contains more than nine-tenths of water, with some solid matters, chiefly urea, the basis of uric acid, and some salts.

These secretions continue at all times, while we are in health, even during our sleep, and are more or less independent of our will.

From this bare outline of Physiology it will be seen how needful it is that all should be clean in person and habits, careful in the kind of food eaten, and in the avoidance of exposure to draughts when heated.

The illnesses of young persons are most frequently caused by want of attention to the kind of food they eat, and by needless exposure.

QUESTIONS ON CHAPTER XLIV.

In what state are the arteries after death?
Describe their construction, and the consequence of it.
How do the veins differ from the arteries?
What is said of the capillaries?
What is the pericardium, and what does it enclose?
Name the principal secretions.
How is the saliva secreted? What is its use?
What is meant by excretion?
By what means does the perspiration escape?
Of what does the perspiration consist?
What are said to promote this secretion?
What are the chief glandular secretions?
By what glands are tears secreted? For what purpose?
What is the chief use of the saliva?
By what gland is the bile secreted?
What are the properties and uses of the bile?
What other fluid assists in digestion?
What office is performed by the kidneys?
Of what is the urine composed?
By what other secretion are the kidneys affected?

ZOOLOGY.

CHAPTER XLV.

ZOOLOGY, (from *zoon*, an animal, and *logos*, a discourse), is a history of the animal kingdom, from man to the lowest subject in the scale of animated creatures, including every creature which is known to have *life, feeling,* and *motion.*

The vast number of animals already known has been classified by various naturalists, from Aristotle downwards, and the classification has been rendered most complete by Cuvier in his Animal Kingdom.

Classification.—It has been found convenient in natural history to arrange animals and plants, minerals, &c., in divisions, classes, orders, genera, and species, in accordance with certain features which are similar in each, and by this means we are better able to remember the various individuals.

All the animal kingdom is thus arranged in four great divisions or sub-kingdoms, beginning from the lowest grade, viz.: *Radiata,* or rayed animals; *Mollusca,* or pulpy animals; *Articulata,* or jointed animals; and *Vertebrata,* or backboned animals.

I. **Radiata** (from *radius,* a ray).—This sub-kingdom is divided into *five* classes, all of which are found in water, and are very simple in structure. They are called Infusoria, Zoophytes, Acalephæ, Entozoa, and Echinodermata.

Infusoria (from *infusor*, I pour in) consists of animalcules, usually found in stagnant water, and so small that they cannot be seen by the unassisted eye, and their habits and formation can only be watched by powerful magnifiers. They were called Infusoria, because all liquids in which either animal or vegetable matter has been steeped or infused, quickly abound with creatures of this class, of various and curious shapes.

They are divided into two groups, one of which has soft bodies, and the other is covered with a shell, which, small as they are, help by their countless millions to swell the great mountains of the cretaceous or chalky formation.

Zoophytes (from *zoon*, an animal, and *phytos*, a plant). These are so called on account of their resemblance to plants. They are very simple in construction, and have this peculiarity, that if cut in two, each separate piece becomes a perfect animal, so that from one specimen many others may be obtained by such divisions.

Examples: the sponge, hydra, coral insects, the actinia, or sea-anemone, and others.

Acalephæ (from *acalepha*, a nettle) are so called because the creatures which compose the order have a power of stinging in the fringe-like tentacula or arms with which they are furnished. While living and in the sea, they look like masses of clear jelly, with the fringe beneath, but out of water they quickly lose their power. Like the sea-anemone, the acalephæ are common on our coasts.

Entozoa (from *entos*, within, and *zoon*, an animal) includes those creatures which live within and prey upon the bodies of other animals. They are usually known as intestinal worms, because one kind of them is sure to be found in the intestines of animals, and

ZOOLOGY.

curiously enough the species found in the ox family is quite different from those found in the horse.

One of these, called the fluke, inhabits the liver of sheep, while several kinds, among others the tapeworm, infest the intestines of human beings. We have no certain knowledge of the means by which they are introduced into the bodies, but most likely it is with raw or unripe food, or food that has become partly decomposed.

Echinodermata, or spiny-skinned (from *echinus,* a hedgehog, and *dermus,* the skin). This class includes the various kinds of star-fish, sea-urchins, and other rough or spiny-skinned rayed creatures.

These have a faculty, like the polypi, of reproducing an arm or ray which has been broken off, and one, the brittle star-fish, will throw off its arms when pursued or alarmed.

The sea-urchin has numerous spines, which help it to move from place to place on the ground.

These, in common with all others of the radiata, have no heart, no circulation, and no distinction of sex.

QUESTIONS ON CHAPTER XLV.

What is zoology? Name some eminent naturalists.
What does the animal kingdom include?
In how many great divisions is it arranged?
Give the meaning of mollusca and vertebrata.
How are the radiata sub-divided?
What are included in the infusoria?
Why are zoophytes so called? Some examples.
Name some peculiarities of the zoophytes.
Where are the acalephæ found?
What are the entozoa? Why so called?
What is said about the horse and the ox?
Where is the fluke found? And the tape-worm?
What is meant by echinodermata?
What creatures are included in it?
What is peculiar to them?
Of what are all the radiata deficient?
Give an example of each radiate order.

CHAPTER XLVI.

SUB-KINGDOM MOLLUSCA.

Mollusca, or soft animals (from *mollis,* soft), are so called because they have no jointed skeleton, nor any connected nervous system, the nerves being spread about in knots or ganglia, and the bodies soft like the snail, oyster, cuttle-fish, nautilus, &c.

They may be conveniently arranged in three classes, ACEPHALA, GASTEROPODA, and CEPHALOPODA.

The **Acephala,** or headless animals (from *a,* without, and *kephale,* the head), include all those shell-fish that are commonly eaten, as the oyster, muscle, cockle, and clam, all inhabitants of water.

The **Gasteropoda,** or belly-footed (from *gaster,* the stomach, and *pous,* the foot), are always found in a one-valve, or univalve shell, and include the slug, snail, and the univalve shell-fish, all of which have the power of increasing and repairing their own shell.

The **Cephalopoda,** or foot-headed animals (from *kephale,* the head). They have long fleshy arms or appendages, which are generally attached to or surround the head. The body in most of these resembles a bag or pouch, to the open end of which the appendages are attached, as in the cuttle-fish, nautilus, &c.

These are common in all seas, but are largest within the tropical, where they are strong enough to seize animals as large as dogs in their long and strong tentacles. They swim backwards, and with their head either upward or downward. They have two large eyes, and are the only molluscs which seem to have a brain in a gristly box.

ZOOLOGY. 157

The cuttle-fish contains a black liquid, which it discharges into the water when disturbed, and which is said to be used in making Indian ink. The fossils Bellerophon, Ammonite, and others belonged to this class.

The paper nautilus is so called because its shell is very thin and light, and it has two tentacles thinner than the others, which can be spread out and hoisted to serve as sails.

SUB-KINGDOM ARTICULATA.

Articulata, or jointed animals (from *articulus*, a joint), includes the four classes, *Annelides, Crustacea, Arachnidæ,* and *Insecta*.

The class **Annelides**, or **Annulata**, ringed creatures, (from *annulus*, a ring) includes all the worm tribe, which are the only animals that have red blood, without being vertebrate or back-boned animals. The body is divided into rings like that of the common worm, with rows of protuberances along the sides of the body.

With the exception of the earth-worm, all the annelides live in water, and breathe through gills, some of which are inside the body, and some on plume-like branches, springing from the head.

Examples—Earth-worm, leech, sea-worm, and the serpula, which is often found in hollow tubes fixed to other sea animals.

The class **Crustacea**, or crust-covered animals (from *crusta*, a shell), includes all the invertebrate creatures which are covered with a hard shell or skin, and nearly all of which have the power of renewing or reproducing a limb if broken off.

Familiar examples of this class are the lobster, crab, craw-fish, prawn, and shrimp, the wood-louse, sand-hopper, and others. The fossil *Trilobite* is also one of this class.

Class **Arachnidæ**, or spider-like animals (from *arachne*, a spider), were formerly classed as insects. They include the various kinds of spider and scorpion, all of which breathe through lungs, and the various mites and ticks, which breathe through air pipes placed on each side of the belly. They have from six to eight eyes, and always eight legs.

There are numerous varieties of spider, some of which are very large, and all of which are carnivorous, or flesh-eaters, and which take their prey in nets, except the mason spider, which constructs a small house, with a trap lid and hinge, lined with a kind of silk. Some dig galleries or pits, and lie at the bottom to entrap the unwary creatures which may attempt to cross, and fall to the bottom.

Scorpions are only found in hot climates, are covered with a hard case, and have a venomous sting in their tails, with which they sting to death locusts, beetles, and other creatures on which they feed.

Mites are usually very small, like the cheese-mites, and similar creatures are found in the skin of animals, which are suffering from cutaneous or skin diseases. Of the same species are the troublesome tick, which is found on sheep and cattle, and which are called *parasitical*, because they hang on them, and suck their blood.

The harvest spider is placed in this order, because it breathes through air-pipes, which differ from the lungs of the other spiders.

Class **Insecta**, or insects (from *inseco*, to cut), includes many orders composed of creatures of great interest, but too numerous to be particularised here.

They nearly all undergo three changes before arriving at the perfect state, which those who have kept silk-worms will readily understand, as the nature of the change is the same in all.

The class is divided into nine orders, which are distinguished by the number or form of the wings. They have all not less than six legs, and numerous eyes.

There are the **Coleoptera**, or sheath-winged, including the beetle, death-watch, cockchafer, and others whose wings are covered with a kind of shell.

Orthoptera, or straight-winged, as the grasshopper and locust.

Neuroptera, or nerve-winged, to which the dragon-fly, May-fly, and white ant belong.

Hymenoptera, or membrane-winged, such as the ant and bee.

Hemoptera, or half-winged, as the aphis and fire-fly.

Heteroptera, or various-winged, as the common bug and water-scorpion.

Lepidoptera, or scale-winged, which includes all the moths and butterflies.

Diptera, or two-winged, as the gnat and common fly.

Aptera, or wingless insects, such as the flea.

In all these names of orders, part of the name is from *pteron*, a wing.

Insects abound in countless millions, especially in hot countries, and some of them perform the work of purifiers, by devouring the animal matter which, if allowed to remain, would be the source of pestilence. They will clean the flesh from the skeleton of a dead animal, as neatly as though it had been scraped with a knife.

QUESTIONS ON CHAPTER XLVI.

What are included in the sub-kingdom mollusca?
Why are they so called? What are the acephala?
Give some examples of the acephala.
What is peculiar to the gasteropoda?
What remarkable power have they?
Describe the cephalopoda. Some examples.
What is remarkable about the cuttle-fish?
Why is the nautilus so called?
How is the sub-kingdom articulata divided?
What are the annelidæ? Where do they live?
To what class does the lobster belong?
What remarkable power have the crustaceæ?
Give some examples of arachnidæ.
What is peculiar in their structure?
Which of them are called parasitical? Why?
Why is the class insecta remarkable?
What are the coleoptera? What is pteron?
To which orders do the bee, moth, and flea belong?
Give examples of diptera, neuroptera, heteroptera.
What great duty do the insect tribes perform?

CHAPTER XLVII.
SUB-DIVISION VERTEBRATA.

The vertebrate animals are divided into four great classes, fishes, reptiles, birds, and mammalia, or *Pisces, Reptilia, Aves,* and *Mammalia.*

These have a backbone containing the spinal marrow, the chief centre or trunk of the nervous system, which is the foundation of the division.

Class **Pisces**, or fishes, (from *piscis*, a fish) all live in water, and are produced from eggs or spawn. They breathe through gills, and are cold-blooded, and have an air-bladder. They are arranged in two orders, Cartilaginous, or gristly, and Osseous, or bony, according to the nature of the vertebra.

The **Cartilaginous** tribes have the skeleton in one piece, and are either fix-gilled, as in the shark or lamprey, or free-gilled, as in the sturgeon and skate.

The **Bony** fish form four orders, according to the structure of the fins: the *Thorny*-rayed, as the perch, flying-fish, and gurnards; the *Soft*-rayed, as the carp, salmon, and herrings, cod-fish, flounder, and other flat fish, eels of all kinds; the *Tufted-gilled*, as the sea-horse and pike-fish; and the *Plectognathi*, or cheek-joined, such as the saw-fish and globe-fish, which have the cheeks or cheek-bones in one piece.

The form of all these creatures is best adapted for moving in the water, and they furnish many instances of design.

All the flat-fish have the eyes on the upper side, and that side is of a darker colour than the other, so as to be less easily seen by others which prey upon them.

All are nearly of the same specific gravity as water, so that a slight inflation or pressure on the air-bladder, will cause them to rise or fall in the water.

Class **Reptilia**, or reptiles, (from *repo*, to creep). These are so called from their creeping mode of walking; they are cold-blooded, and slow in respiration or breathing.

Reptiles are oviparous, that is, produced from eggs or spawn, grow slowly, sleep much in cold climates, and usually live to a great age.

Like the lowest class Radiata, many of them have the faculty of reproducing limbs if broken off, or an eye if taken out, and are remarkably tenacious of life. Some, as the frog for example, will live a long time after the brain has been removed from the head, while a tortoise will live for weeks without any head at all.

Reptiles are divided into four very distinct orders: *Chelonia, Sauria, Ophidia,* and *Batrachia.*

The order **Chelonia** (from *chelys*, a tortoise) includes

14—3

all the tortoise and turtle tribes, which are enveloped in a hard shell composed of two plates, often of great strength. They have no teeth, but the jaws are so hard, that they are able to bite through shrubs and the toughest vegetables.

Most of the tortoises proper are found on land, though a few inhabit both land and water. The turtle, or marine tortoise, lives in the sea, but comes frequently to shore to lay its eggs in the sand, and to feed.

One species, the hawk's-bill turtle, furnishes the fine tortoise-shell which is so valuable in commerce, and which is separated from the creature by a cruel roasting process, while the animal is alive. The turtle of the Galapagos islands, and of other tropical parts, affords excellent food, and a good oil is obtained from its flesh.

The order **Sauria**, or lizard-like animals, (from *saura*, a lizard) includes the crocodiles, chameleons, geckos, which have the singular power of adhering by their peculiar sucker-like feet to the ceiling of a room; iguanas, which are remarkable for their large size, and a leathery membrane, which enables some of them to make long leaps from tree to tree; the true lizards, such as have a loose forked tongue, and five toes on all the feet; and the skinks, which are more snake-like in form than the others. The last are able to burrow into the sand with great rapidity.

The order **Ophidia**, or serpent-like, (from *ophis*, a serpent) includes all the snake and serpent tribes. Nearly all are destitute of limbs, though a few have little bones beneath the skin, which look like the rudiments of limbs.

In all the true serpents the ribs pass completely round the body, the jaws are movable and easily separated, so that these creatures can swallow others of larger size than themselves.

ZOOLOGY. 163

Few of them in comparison are poisonous. Those which are so, are furnished with two fangs, or long hollow teeth, and at the base of each hollow there is a spongy gland, which when pressed yields a poisonous fluid, that passes down the groove of the teeth into the wound made by the serpent.

These fangs, when not in use, fall backwards to the roof of the mouth, so as not to be in the way while food is being swallowed.

The order Batrachia, or frog-like animals, (from *batrachos*, a frog) include frogs, toads, the salamander, proteus, and siren. The frog tribes pass through some curious changes before attaining their true shape. The spawn or egg becomes a tadpole, which is furnished with gills for breathing purposes, and which fall off when the animal changes from the tadpole into a frog or toad. They feed chiefly on insects.

The salamander, proteus, and siren, are in form more like the saurians, but they undergo similar changes to the other Batrachians. They are found chiefly in mines, or other subterranean passages. Some gigantic fossils of the salamander have been found in Germany, and parts of them in England.

QUESTIONS ON CHAPTER XLVII.

How is the sub-kingdom vertebrata divided?
What is said of all these classes?
What is said of the blood of fish and reptiles?
How are they produced? How divided?
What are cartilaginous fish? An example.
How are bony fishes arranged?
Give an example of each of the four orders.
What is said about the form of fishes?
What are the orders of the class reptilia?
What is included in the chelonia?
Whence is tortoise-shell obtained?
What are included in the order sauria?
Name some peculiarity of the gecko and iguana.

To what order do the serpents belong?
Describe the mode of poisoning in serpents.
What are included in the order batrachia?
Through what changes does the frog pass?
Why is the salamander placed with the batrachians?

CHAPTER XLVIII.

CLASS AVES.

This class (from *avis*, a bird) are all oviparous and warm-blooded, with a complete double circulation, are nearly all covered with feathers, and are bipeds, or two-footed.

Birds have air cells in their bones, which communicate with the lungs, &c., by which their specific gravity or weight is lessened, and the lungs are fastened to the cavity of the chest.

As they have no teeth, they are furnished with a powerful muscular apparatus called the gizzard, which in some of the class is so powerful, as to grind down metallic substances.

In all birds the eye is formed for distant sight, and in some it is covered with a third eyelid, called the nictitating membrane; this is drawn forward by the bird at will, and forms a sort of screen to the eye, by means of which the sight of eagles, hawks, and other birds of prey is preserved and strengthened.

Birds are divided into seven orders, viz.: *Accipitres, Passeres, Scansores, Gallinaceæ, Grallæ, Cursores,* and *Palmipedes.*

Accipitres, or hawk-like, (from *accipiter*, a hawk). These are also called raptores, or plundering birds, because they are all birds of prey.

They are distinguished by their hooked beak and claws, and have strong muscular powers of wing.

They are divided into *Diurnal* birds, such as the eagle, vulture, hawk, falcon, buzzard, secretary, and kite, which fly by day, and the *Nocturnal* raptores, as the owl.

The nocturnal birds are singularly adapted for silent hunting by night. Their eyes are large, so that they receive every ray, and the feathers are so soft and downy, that they make scarcely any sound in flying.

All the accipitres are solitary, and seldom lay more than two eggs. Were it otherwise, they would multiply, so as to destroy too many of the smaller creatures on which they prey.

The buzzard and vulture are useful in tropical climates to clear away the dead bodies of animals and other garbage. The rest feed only on living prey. In Egypt the common vulture acts as a useful scavenger, hopping about the streets feeding on garbage. On this account it is protected by law, like the stork in Holland, which serves a similar purpose.

Order **Passeres**, or sparrow-like birds, (from *passer*, a sparrow) are also called insessores, or perching birds. This order comprises numerous varieties, which differ in many respects, but are all perchers.

They are **Fissirostres**, or wide-mouthed birds, such as the night-jar, swallow, kingfisher, and others, which catch their insect prey as they fly with open mouths; all these have short legs, and generally weak feet.

Dentirostres, or tooth-billed birds, as the butcher-bird, thrush, fly-catchers, and warblers. These have a notch or tooth on each side of the upper mandible of the bill.

Conirostres, or cone-billed birds, as the starling, sparrow, lark, crow, various finches and others, which feed chiefly on grain.

Tennirostres, or thin-billed birds, as the humming-bird, sun-bird, and hoopoes. These frequently have the tongue divided at the tip, to aid them in sucking honey from flowers.

Some naturalists have arranged them differently, but the above is more convenient than any other.

Scansores, or climbers, compose the third order. Instead of having three toes in front and one behind, as is the rule, the climbers have two toes in front and two behind, by which they are enabled to grasp more easily whatever they wish to climb or hold. This includes the parrot, toucan, wood-pecker, tree-creeper, and cuckoo.

Gallinaceæ, or fowl-like (from *gallina*, a hen). This order comprises the common fowl in all its varieties, pheasant, grouse, turkeys, pea-fowl, and the pigeons.

These are all useful to man, and some varieties, like the cud-chewing animals, are found wherever man is found. They are gregarious, lay many eggs, and are granivorous, or grain-devouring. They are more capable in walking and running than in flying, and because of their habit of scraping or scratching the ground, they are sometimes called **Rasores**, or scrapers. Many of the most valuable have been reclaimed from the wild state.

The **Columbæ**, or pigeon family, are sometimes placed in a distinct order, because they differ somewhat from the rest of the Gallinaceæ. They breed in pairs, lay only two eggs for a hatching, and feed their young ones from their own crop, in which they soften the grain to prepare it for them.

Grallatoræ, or **Grallæ**, stilted birds (from *grallæ*, stilts), compose the fifth order. These are all wading birds, long-legged and long-necked; they all usually have the bill long, and fitted for searching the waters for food.

ZOOLOGY.

The chief examples of this order are the stork, heron, lapwing, crane, snipe, and other similar long-legged birds.

They feed chiefly on small fish or reptiles, and in Holland the stork is protected, as being very valuable in destroying reptiles and noxious insects.

Cursores, or runners. By some naturalists the ostrich, emeu, cassowary, apteryx, and the extinct dodo, are classed with the Grallæ, but they are more properly called Cursores, or runners.

They do not fly, for their wings are not sufficiently long, though they make use of them in running.

In some of them, as the emeu of Australia, and the apteryx of New Zealand, the feathers look more like hair, and their food is obtained in an entirely different mode from that of the Grallæ.

Palmipedæ, or web-footed birds (from *palma*, a palm), includes all the really web-footed swimming birds. This order is sometimes called also **Natatores**, or swimmers.

It includes all the flat-bills, such as ducks, geese, and swans; the completely webbed, as the pelican; the long-winged, as the albatross, gull, tern, and petrels; and the short-winged, as the penguin, auk, puffin, and various divers.

Some of these, as the petrel and albatross, live entirely on the sea, while others live partly on land, and inhabit the rocky coasts and islands of distant parts of the world.

QUESTIONS ON CHAPTER XLVIII.

Name the peculiarities of the class aves.
What have they instead of teeth?
What is the nictitating membrane?
How is the class aves divided?
Which belong to the diurnal accipitres?

How are the nocturnal birds adapted for hunting?
How are the vulture tribe found useful?
How are the sparrow-like birds arranged?
To what sub-order do the swallow and thrush belong
What are those called which feed on grain?
What is remarkable in the climbers?
Which order is most useful to man?
Give some description of the gallinæ?
Why are they sometimes called rasores?
In what do pigeons differ from the rest?
What are included in the grallatoræ? And why?
Where is the stork found very useful? Why?
To what order does the ostrich belong?
What is curious in the emeu and apteryx?
In what order are the flat-billed birds?
Name some differences in the birds of this order.

CHAPTER XLIX.

THE CLASS MAMMALIA.

The class **Mammalia** (from *mamma*, a breast or pap), includes all those animals which have a complete double circulation of warm blood, and are viviparous, that is, which produce their young alive, and which moreover suckle them with milk, until able to provide food for themselves.

This class is divided into nine orders, viz. : *Bimana, Quadrumana, Carnaria, Marsupialia, Rodentia, Edentata, Pachydermata, Ruminantia,* and *Cetacea.*

Order **Bimana**. This order (from *bis*, two, and *manus*, a hand,) includes man only, who alone of all creatures can rightly be called two-handed. As we shall enter minutely into the structure of man in Human Physiology, further mention of it here is omitted.

Order **Quadrumana** (from *quatuor*, four, and *manus*,

a hand). In this section all the ape, monkey, and lemur tribes are included. They are called four-handed, because what is usually looked upon as the hinder foot, is really a hand.

The thumb is placed opposite the fingers, and not beside them in all the four limbs, and the fingers are long and flexible, so that the quadrumana can grasp as readily with one hand as the other. As they swing about much on trees in their search after food, this arrangement is a great convenience to them.

Some writers have shewn a desire to prove that this order is nearly identical with the order Bimana.

In spite of their intelligence, however, the structure of the skeleton in the ape, as well as the rest, shows that they are quite distinct from man.

Apes are destitute of tails, baboons have short ones, and monkeys generally long ones, and frequently prehensile, that is they are able to use them as a hand to grasp the branch of a tree, &c.

The lemur is found in Madagascar, and is nocturnal; it is also more slender and delicate than others of the order. The monkey tribe are nearly all fruit-eaters, and are themselves roasted and eaten with relish in South America.

Order **Carnaria** (from *carnarius*, a butcher), includes all the killing animals. They may be known by the formation of the teeth, which are only formed for cutting and tearing food, and not for grinding it, as in the vegetable and grain eaters.

This is a very extensive order, and includes very various creatures, and it is usually divided into five great families, which are themselves so large, that some naturalists give them a separate order.

They are *Cheiroptera*, or hand-winged; *Insectivora*, or insect eaters; *Plantigrada*, or foot-walkers; *Digitigrada*, or toe-walkers; and *Amphibia*. The last three compose the true carnivora, or flesh-devourers.

15

Cheiroptera (from *cheir*, a hand, and *pteron*, a wing), includes the bat tribes, which are found in all temperate and tropical climates. They are distinct from all other creatures, in having a body similar to that of a mouse, while the limbs are connected by a leather-like membrane, which when spread, serves them as wings.

In the East Indian Islands there are bats of large size, which are fruit-eaters, and in South America others which suck the blood of animals, and are called vampire bats.

Insectivora (from *insecta*, an insect, and *voro*, I devour), includes the mole, hedgehog, shrew, muskrat and tenrec, all of which are insect-eaters.

Plantigrada (from *planta*, the sole of the foot, and *gradus*, a step), include all the bear tribes, the racoon, glutton, ratel, and badger, which plant the whole foot on the ground; the footstep of the bear in the snow is easily mistaken for that of a man.

Digitigrada (from *digitus*, a finger), include all those that walk on the toes, viz.: the dog tribe, with the wolf, fox; the cat tribe, with the lion, tiger, jaguar, leopard, and the common cat, weasel, polecat, and many others.

The cat tribe have their claws in a sheath, which keeps them always sharp, with a soft ball under each toe, and one under the foot. Nearly all of them take their prey by springing upon it.

The dog tribe have the claws always unsheathed, and hunt their prey down in packs.

Amphibia (from *amphi*, both, and *bios*, life), includes all the varieties of seal.

Order **Marsupialia** (from *marsupium*, a pouch), includes the kangaroo, opossum, wombat, ornithorynchus, and others, that are furnished with a belly-pouch, in which the young ones are carried and suckled until

ZOOLOGY. 171

strong enough to take care of themselves. Except the opossum, they are all peculiar to Australia. Most remarkable of them all is the ornithorynchus, or duck-billed platypus, which has a body somewhat like a beaver, though smaller, with the bill of a duck, webbed feet, and spur on the heel. The kangaroo is herbivorous, but the others are insectivorous.

Order **Rodentia** (from *rodo*, I gnaw), comprises those animals which have two long front incisors, or cutting teeth, in each jaw, and no tearing teeth, such as the rat, rabbit, squirrel, beaver, porcupine, and many others.

All these have greater strength in the hinder than in the fore legs, and some, as the squirrel and beaver, make use of their fore-paws as hands. All are herbivorous, and many feed on roots.

QUESTIONS ON CHAPTER XLIX.

How does class mammalia differ from the other classes?
How is it arranged? Where is man placed?
What are included in the quadrumana? And why?
How do apes, monkeys, and baboons differ?
Where is the lemur found?
In what order are the chief beasts of prey?
How may they be known?
Into how many great families are they divided?
What are the cheiroptera? Why so called?
Where are the bat tribe found very large?
To what family do the mole and hedgehog belong?
What is meant by plantigrada?
What animals belong to the digitigrada?
What is remarkable in the cat tribe?
What is meant by amphibia?
What is peculiar to marsupialia?
Where are they found? With what exception?
Which is most remarkable of the marsupials?
In what order are rabbits included? Why?
What is said of the legs of the rodentia?

CHAPTER L.

MAMMALIA—*continued.*

Order **Edentata** (from *e*, without, and *dens*, a tooth), includes those animals which have no front teeth, that is, no incisors, and sometimes no tearing teeth, as the sloth, armadillo, and ant-eater.

The sloth feeds on the leaves and small branches of trees, and is curiously slow in its movements; the others feed chiefly on ant-hills and putrid flesh. They are found only in South America, where they are eaten by the native Indians.

The armadillo is covered with a hard, striped, shell-like armour, and like the ant-eater, is entirely destitute of teeth.

The ant-eaters obtain their food by laying their long slimy tongue in the path of the ants, which are caught by the slime, and eaten by the ant-eater. Several gigantic fossil specimens of this order have been found, which have been called the megatherium, or great beast, and the megalonyx. These are of the sloth tribe.

Order **Pachydermata** (from *pachus*, thick, and *dermis*, the skin), comprises a great number of animals, many of which are very large, as the elephant, rhinoceros, hippopotamus, taper, common hog, horse, and all the zebra and ass tribes. They have all a very thick skin.

The elephant differs from the rest in having a hollow trunk or proboscis, which no other extant animal has. The horse, ass, and zebra families are distinguished as *solipeda*, or solid-footed, because they differ from the rest in having the hoof undivided.

None of this order chew the cud, though they are

all granivorous or herbivorous, and some, as the hippopotamus, are a great scourge to the rice and corn fields in countries where they abound.

Order **Ruminantia** (from *rumino*, to chew the cud). This order is one of the most valuable to man, as all the animals which compose it are useful in various ways. It includes all those which divide the hoof and chew the cud, or ruminate.

Except the camel, they have no front teeth or *incisors* in the upper jaw, but they are provided with four stomachs for the purpose of completely digesting their vegetable food.

They have the power of returning the food in small parcels from the second or honey-combed stomach to the mouth, to be chewed a second time.

It contains two families, the horned and the hornless; the latter contains the camels, musk-deer, llama, and vicugna.

The horned ruminants comprise oxen, deer, sheep, goats, antelopes, and giraffes. The antelopes differ from the deer in having hollow horns. Many of these labour for man while living, and supply him with food and clothing when dead.

The foot of the camel is unlike that of other ruminants, as it is large, broad, and spongy, so that it does not sink into the sand.

The foot of the reindeer opens widely to prevent that animal sinking into the snow over which it passes. Wherever man can live, there some ruminating animal is to be found.

Order **Cetacea** (from *cetus*, a whale). The ninth and last order of the class Mammalia, includes the whale, narwhal, dolphin, porpoise, and other whale-like animals.

They have more blood than other animals, as well as a larger amount of fat under the skin, which makes them lighter in the water, and is thought to enable them better to bear pressure.

They have also the power of forcing water through a hole in the upper part of the head, on which account they are sometimes called blowers.

By common error whales have been called fish, but as fish are cold-blooded, and produced from eggs, while the Cetacea are warm-blooded, and suckle their young, the distinction is plain.

Cuvier placed among the Cetaceans the manati and dugong, two vegetable eaters, which are found in the eastern seas and in the Atlantic. They are eaten by the Malays, and have very much the appearance of small whales, but are without the orifice in the head, and have less blubber or fat.

From the above outline it may be seen that the range of Zoology is very extensive, and few persons can give attention to all its branches. It is divided into various branches, each of which affords abundant materials for observation.

The principal divisions of the science are *Conchology* (from *conke*, a shell), the science of shells and shell-fish; *Entomology* (from *entoma*, an insect), the study of insects; *Ichthyology* (from *ichthus*, a fish), the study of fishes; *Ornithology* (from *ornis*, a bird), the study of birds; and *Zoophytology*, the study of the plant animals.

QUESTIONS ON CHAPTER L.

What is remarkable in the edentata?
What exception is made in the order?
Describe the armadillo and ant-eater.
What great fossils are of this order?
What is meant by pachydermata?
Which are called solipeda? And why?

How does the elephant differ from the others?
Which order is most valuable to man?
Why are they called ruminants?
Which are the hornless ruminants?
How does the camel differ from the others?
What is remarkable in the reindeer?
Where are the ruminants found?
What animals compose the last order?
What is said of the blood and fat of the cetacea?
What error has been common about whales?
Show how they differ from fishes.
What is meant by conchology?
What science describes insect life?
How are the studies of birds and fishes named?
What is zoophytology? Why so called?

INDEX.

	PAGE		PAGE
Acetic acid	90	Carnaria	169
Acrogenæ	56	Cetacea	173
Affinity	60	Chemistry	58
Alchemy	58	Chlorine	67
Alcohol	98	Chronometers	13
Aluminium	74	Circulation	146
Annelides	157	Copernicus	6
Annual motion	10	Copper	77
Antimony	75	Corolliflora	53
Arachnidæ	158	Creosote	90
Arsenic	75	Cryptogamiæ	46
Articulata	157	Crustacea	158
Arteries	149	Cursores	167
Astronomy	4		
Aves	164	Dew	135
		Digestion	141
Batrachia	163	Diurnal motion	8
Bile	143		
Bimana	167	Earth	7
Botany	27	Eclipses	15
Butter	94	Edentata	172
		Electricity	100
Caloric	127	Endogenæ	55
Calyciflora	51	Exogenæ	40
Caoutchouc	95		
Capillaries	149	Firedamp	70
Carbonic acid	69	Fixed stars	26

INDEX.

	PAGE		PAGE
Fixed oils	93	Nitrogen	66
Fossils	123	Nutrition	144
Freezing mixture	135		
		Ophidia	162
Gallinæ	166	Organic chemistry	88
Galvanism	108	Oxalic acid	89
Gases, The	65	Oxygen	65
Geology	115		
Glass	85	Pachydermata	172
Gold	79	Palmipedes	167
Gum	97	Pancreas	151
Gun cotton	99	Passeres	165
Gutta Percha	96	Perspiration	150
		Pisces	160
Heat	127	Planetoids	7
Heart, The	146	Plants	28
Hydrogen	66	Plant food	32
		Plaster of Paris	76
Ice	132	Platinum	83
Insectæ	158	Porcelain	85
Iodine	72	Potassium	84
Iron	79	Physiology	137
Jupiter	23	Quadrumana	168
Latitude	14	Radiata	153
Lead	81	Reptilia	161
Lightning	106	Rocks	116
Longitude	13	Rodentia	171
		Ruminantia	173
Magnesium	81		
Magnetism	112	Safety lamp	71
Mammalia	168	Saliva	141
Mars	22	Saturn	24
Marsupialia	170	Sauria	162
Mercury	21	Scansores	166
Metals	74	Seasons	11
Mollusca	156	Secretions	149
Monochlamydeæ	54	Silver	86
Moon, The	17	Skin, The	139
Mosses	47	Soap	94
Muscles, The	139	Solar System	5
		Starch	97
Natural system	44	Stomach, The	141
Neptune	25		

INDEX.

	PAGE		PAGE
Sugar	97	Urea	89
Sulphur	73	Urine	151
Tears	150	Venus	22
Teeth	140	Ventilation	129
Tides, The	14	Vertebrata	160
Tin	86	Vulcanite	95
Thalamifloræ	49		
Thermometer	133	Zinc	87
		Zoology	153
Uranus	25	Zoophytes	154

THE END.

BILLING, PRINTER, GUILDFORD.

WORKS BY
JOSEPH FERNANDEZ, LL.D.

Henry's Outlines of English History.
A Complete Synopsis of the National History and Constitutional Progress, with GENEALOGICAL TABLES OF EACH FAMILY, 1500 Questions, and Chronological Tables of Principal Events.
Sixth Edition. 228 pages, foolscap 8vo. Price 2*s.*

Henry's Junior Dictation Lessons.
In HENRY'S JUNIOR DICTATION LESSONS, the words which are to be learned are printed in distinct type at the head of each Lesson, as it is desirable that the learner should *see* words, the sounds of which are often so different from their appearance.

It contains 174 Lessons, or more than 2000 words, from *one* to *three* syllables, special attention being given to those which are most likely to be mis-spelt or mis-applied.
Just Published. 174 pages, foolscap 8vo. Price 1*s.* 6*d.*

Henry's Dictation Lessons.
These Lessons, on about 5000 words, are arranged in Three Parts :—
1. Consisting of Ninety Lessons on *Words Alphabetically Arranged.*
2. Thirty-two Special Lessons on *Peculiar Forms of Spelling.*
3. A recapitulatory series, containing Lessons on all the most difficult words, arranged according to *Accent and Vowel Sounds.* The sentences are framed so as to convey *some fact worth knowing, or some moral truth.*
Third Edition. 174 pages, foolscap 8vo. Price 1*s.* 6*d.*

Henry's School Geography.
HENRY'S SCHOOL GEOGRAPHY has been written with the view of providing a book which would be sufficiently full for ordinary school classes without trenching on the province of those that are more exhaustive, and special attention has been paid to the geography of the United Kingdom, the British Colonies, the States of America, and the leading states of Europe.

More than 3000 questions on the text have been inserted at the end of the various sections.
Third Edition. 226 pages, foolscap 8vo. Price 2*s.*

LONDON:
CHARLES BEAN, 81, NEW NORTH ROAD, HOXTON.

WORKS BY

REV. J. ROBERTSON.

Half-Hour Examination Papers for Daily Use.
These papers are intended to form a useful adjunct to a pupil's daily study. The subjects are such as are commonly taught in Middle-Class Schools, and form an excellent preparation for any of the now prevalent Examinations.
Price 1s. 6d. 144 pages. *New Edition—Third Thousand.*

Answers to Half-Hour Examination Papers.
Price 3s. 6d.

Early Latin Exercises for the Use of Young Beginners. The object of this Exercise Book is to ground young beginners in the declensions of substantives and adjectives. The Exercises have been prepared with a view to provide the pupil with numerous examples, and are so arranged as to thoroughly test his learning.
Price 1s. 6d. 88 pages. *Second Edition.*

Examination Essentials.
Part 1, price 2s. 6d. Part 2, price 4s. 6d.

Daily Exercises in Arithmetic.
"Practice makes perfect." These Exercises are adapted for daily use, and are intended to *perfect* the pupil's knowledge of the various rules of Arithmetic. They contain 1295 questions.
Price 1s. 6d. 144 pages. *Third Thousand.*

Answers to Daily Exercises in Arithmetic.
Price 1s. 6d.

Gospel Questions, or a Course of Lessons on our *Lord's Personal History.* To which are added, HINTS AND AIDS FOR ANSWERING GOSPEL QUESTIONS.
Price 2s.

LONDON:
CHARLES BEAN, 81, NEW NORTH ROAD, HOXTON.

[P. T. O.

www.ingramcontent.com/pod-product-compliance
Lightning Source LLC
Chambersburg PA
CBHW020251170426

43202CB00008B/313